Anytime Weather Everywhere

A book of AWEsome weather experiments and investigations for early childhood level scientists

(also for use by teachers, parents, naturalists, and others interested in the science of meteorology)

by H. Michael Mogil and Barbara G. Levine

HOW THE WEATHERWORKS
1522 Baylor Avenue
Rockville, Maryland 20850

Anytime Weather Everywhere is one of several supplemental curriculum materials for studying about the weather which are available from HOW THE WEATHERWORKS. Please write to obtain the latest catalog describing cloud post cards, cloud flash cards, weather posters, and other materials.

Anytime Weather Everywhere

FIRST EDITION
FIRST PRINTING

© 1996 H. Michael Mogil and Barbara G. Levine
HOW THE WEATHERWORKS
1522 Baylor Avenue
Rockville, MD 20850

Printed in the United States of America. All rights reserved. Data and activity sheets may be reproduced for classroom use; otherwise, no portion of this book (or the accompanying cloud flash cards) may be reproduced in any form or by any means without the express written permission of the authors.

Due to the nature of scientific inquiry and the numerous warning messages noted throughout this book, the publisher and the authors take no responsibility for the use of any materials or methods described in this book, nor for the products thereof.

ISBN 1-887-01349-0

Table of Contents

Introduction 1

Sky 4

1	Take a Hike!	4
2	Take a Colorful Hike!	5
3	I'm a Weather Reporter	8
4	Cloud Watching	13
5	Creating Clouds	16
6	Creating Clouds (cotton balls)	18
7	Creating Clouds (in a bottle)	19
8	I'm In a Fog	21
9	A Cloud and Sky Watcher Window	22
10	Sky Diary	26
11	Raining Cotton Balls	27
12	Keeping Raindrops Up!	30
13	The Great Raindrop - Snowflake Race!	31

Water 32

14	Sky Inventory	32
15	Water Cycle Web	33
16	Around and Around the Water Cycle	34
17	It's Raining Without Raining	35
18	Now It's Really Raining	36
19	A Mushroom in the Rain	37
20	Playing with Raindrops	38
21	Surface Tension	39
22	Fossilized Raindrops (A great activity for outdoors)	42
23	Measuring the Rain	44
24	How Much Water Is In My Snow?	46
25	Going, Going, Gone Once	48
26	Going, Going, Gone Twice	49
27	Ice is Nice	50
28	Icebergs	51
29	Ice to Vapor	52

Wind 53

30	Air, Air, Everywhere	53
31	Up, Up and Away	54
32	Batter Up!	55
33	Balloon Rockets	56
34	Balloon Rocket Race	57
35	Flighty Activities	58
36	Where the Wind Comes From (Building a wind vane)	62
37	Blowing in the Wind	64
38	Air and Water Streams	65
39	Wind Speed Indicator (Building an anemometer)	66
40	Pinwheel Mania	67
41	Make Your Own Wind Machine	68
42	Tornado!!!	69

Sun 71

43	Sun Prints	71
44	Sunbathing	72
45	Shadows	73
46	Shadow Tag	74
47	Take a Shadow Hike	75
48	Cloud Shadows	76
49	Groundhog Day	77
50	Plants	78
51	Sundials	79
52	Day and Night	80
53	A Solar Web	81
54	Rainbows	82
55	The Story of the Rainbow	83
56	Sunshine to You	85

Temperature 86

57	Human Thermometers Aren't So Hot	86
58	Reading a Thermometer	87
59	Expanding Your Understanding of Temperature	91
60	Looking for Warm and Cool Places	92
61	Things Are Heating Up	93

62	**The Heat Is On**	95
63	**Now You See It - Now You Don't**	96
64	**Putting Chemistry on Ice**	98
65	**We Get to Eat an Experiment!!!**	99
66	**Blowing Hot and Cold Bubbles**	101
67	**'Tis the Season**	102

Resources and References 104

Reproducible Masters

COLORS IN NATURE Chart	7
Weather Calendar	12
SKY and CLOUD WATCHER Data Record	25
"RAINING COTTON BALLS" Prediction Data Sheet	29
"SURFACE TENSION" Prediction Data Sheet	41
Measuring Precipitation	45
"FLIGHT DISTANCES" Data Sheet	60
SCIENTIFIC METHOD Data Sheet	61
My Temperature Observations	88
Melting Predictions and Observations	97
"amateur meteorologist" certificate	103

Introduction

This book is about weather. But it is also about how weather can be used to study mathematics, reading, and other disciplines. By using the activities described in this book, you and the children you work with will learn better how to observe, classify, hypothesize, predict and experiment through extensive studies about the weather.

> **Observe**
> collect and record data
> use sight, sound, smell, and touch
> to become aware of your environment
>
> **Classify**
> report on observations
> graph data
> match cloud pictures
>
> **Hypothesize**
> propose explanations
> solve problems
> reason
>
> **Predict**
> make forecasts
> verify forecasts
>
> **Experiment**
> test and compare results
> control variables

This book contains nearly a hundred activities, experiments, investigations, demonstrations and challenges which can be used to create a two to four week weather unit. Select those that best fit the current weather and the interests of the children, as well as any curriculum requirements you may have. Feel free to expand upon what is contained in this book as you become more comfortable working with the concepts.

We encourage you to include weather activities throughout the entire school year, not only when you do a weather unit. One way is to begin a weather calendar at the start of the school year. Continue collecting and recording data on the calendar for at least three months. The calendar, with an appropriate legend (Activity 3), can be used to stimulate classifying, graphing, and comparing activities over several seasons.

Because this book spans several grade/age levels, you may find some sections not appropriate for the developmental level of your particular group of children. For example, in Activity 2, children in grade 2 may be able to easily read and write the color words. Kindergarten children may require additional preparation; they may have to copy the words and color in the word boxes with the color name match.

Although you may already use the following teaching techniques, we mention them just as a reminder. As necessary, cover work surfaces and provide clean up materials, anticipate extra needs (e.g., extension cords); try to do activities that involve drying time earlier in the day; and adapt your planned schedule when the weather does or doesn't cooperate.

We have provided suggested sizes for some materials. If these are unavailable, substitute as necessary. If your school or center does not collect items such as paper towel rolls (Activity 32), we suggest that you ask parents to help provide them.

To make this book even easier to use, each activity contains objectives, a listing of materials required, and appropriate background material. Many of the activities contain one or more progressive activities and suggested discussion questions (in bold text). In some activities or demonstrations, you may need to lead the children through a discussion or brainstorming activity rather than asking specific questions.

All of the activities, experiments, and demonstrations are ideal candidates for science and mathematics journal entries. The questions (bold font) provide prompts for having children describe what they saw and summarizing their understanding of it. You will also find many reproducible master laboratory/data sheets throughout the book. You have permission to use these within your class or to adapt them.

We have also added "LEADER CHALLENGES" as opportunities to extend weather study both in the classroom and into the home. You should always encourage the children to share *all of their experiences* with their families. This will enable them to assume the role of teacher and to reinforce their understanding of newly gained knowledge.

Before you begin studying weather, please familiarize yourself with the list of resources and references found at the end of this book. We have chosen more than a dozen reading and picture books because they fit so well into the activities described in this book. Other references provide technical information which may help you to better understand weather concepts. Also, visit your school's library, a public library, and/or a children's book store to locate other books or resource materials about weather. For current information about weather (locally and in other places) we encourage you to watch the morning and evening television weather reports, watch The Weather Channel™ on your local cable system, and read your local newspaper.

This book comes with a set of nine cloud flash cards for use with Activity 4. We have also produced four highly visual posters which support (but are not required for) activities and concepts discussed in this book. See supplemental resources for information about ordering these.

We have used English units throughout this book because that is how weather data is reported on television and in newspapers. However, we encourage you to introduce children to metric units and/or use dual units, whenever possible.

We cannot over-emphasize the importance of safety in a science setting. Although we have not specifically noted safety issues in each activity, we urge you and your children to be careful whenever involved with science. This includes using common sense when looking at the sky (i.e., don't look directly at the sun), avoiding putting things in the mouth (unless instructed to do so), being careful with supplies and materials (e.g., scissors, glass jars, electricity, and heat), and dressing appropriately for outdoor activities. Be sensitive to food allergies some of your children may have, too. Substitute ingredients for some of the "eating" activities, as needed. With proper adult supervision and "warnings", some demonstrations can involve child volunteers (Activity 12). If you are already science safety conscious, super!! If not, please heed these cautions and help make your children safety minded, too.

We encourage you to incorporate music, poetry, and fine arts into these activities, too. Your local museum art gallery, local or school library, and fine arts specialists can recommend resource materials and/or help you.

Activities have been coded to reflect curriculum focus areas and emphasize safety. Refer to the legend below for further information.

Finally, if you have any questions about the activities, or wish to share some special experiences with us, please write to us at HOW THE WEATHERWORKS, 1522 Baylor Avenue, Rockville, MD 20850. We'd love to hear from you!

Have a wonderful time studying the weather!!!

H. Michael Mogil
meteorologist

Barbara G. Levine
teacher

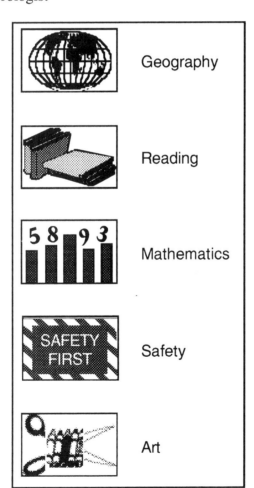

Sky

1. Take a Hike!

Activity

Objectives
- to become aware of your surroundings
- to collect data
- to record your observations
- to graph results

Materials
- none

Weather permitting (or not), dress for the weather and take your children on an observation hike/walk to observe the weather and things in nature. Afterwards, make a list of what the children have seen, smelled, or heard during their hike and how these may or may not be related to weather. This activity can be done in urban, suburban, and rural settings.

Using the list, create a number graph comparing the number of things that were seen with the number of smells and sounds. **Which of the three categories had more?**

Create subsets containing only those sights, smells, and sounds that relate to weather. **Which of the three categories had more?**

Post lists in your classroom for reference.

Repeat the activity on another day and in other seasons. See if the children can add to these lists.

REMEMBER to emphasize to the children that they should not taste things during science study unless they are given specific instructions to do so! Also, they should be told NOT to look directly at the sun.

Suggested reading: *The Listening Walk*

2. Take a Colorful Hike!

Activity

Objectives
- to become aware of your surroundings
- to make predictions
- to collect data
- to record and report your observations
- to graph results

Materials
- 8 1/2" x 11" poster board, markers, crayons, colored stickers, master data sheet

Before taking your hike, have each child make a color chart like the one shown on Page 7; they can write the words or just color in the boxes with the appropriate color. Alternatively, you can make copies of the reproducible master (Page 6) for each child.

We have shown the basic "rainbow" colors, brown, and the basic black-white-grays. You may add other colors, as appropriate, based on factors unique to your local area.

The charts should be done or pasted on 8 1/2" x 11" poster board. The "colors" can be on one side of the poster board, the "grays" on the other. Have each child predict what color(s) they think they will find most often in nature, noting this on the top of their charts. Provide each child a marker, crayon or colored stickers so the child can tally mark his/her charts each time they see a colored nature item. After taking the walk and returning to the classroom, have children count the number of different times they saw a color during their hike. Have children report on their observations. Compare these with their predictions.

Additionally, children can record the **names** of the actual objects found in nature that were different colors (e.g., writing green "grass", red "flowers", and blue "sky").

What color(s) did children see the most? Does this have anything to do with the weather? Do you think the colors would be the same in all seasons? How might they differ? Who looked at the sky? What color was it? What color(s) were the clouds? Repeat this activity in different seasons and compare and contrast results.

COLORS IN NATURE CHART

red	orange	yellow	green	blue	purple	brown	white	black	gray
flower	*flower*	*leaves*	*grass*	*sky*	*flower*	*tree*	*clouds*	*clouds*	*clouds*
leaves	*leaves*	*flower*	*leaves*	*house*		*sand*	*flower*	*birds*	*car*
berries			*moss*	*car*		*patio table*	*fence*	*ants*	*rocks*
				pond		*slide set*	*trash papers*		*bird pigeon?*

COLORS IN NATURE CHART

red	orange	yellow	green	blue	purple	brown	white	black	gray

© 1996...H. Michael Mogil and Barbara G. Levine, 1522 Baylor Avenue, Rockville, MD 20850

3. I'm A Weather Reporter

> **Activity**

Objectives
- to become aware of the weather
- to collect and record data
- to report observations
- to make weather predictions (forecasts)

Materials
- poster board, markers, crayons

Discuss what makes up the weather (clouds, color of the sky, precipitation, sun, temperature, and wind) and why it is important to us. Keep a daily weather calendar. Then make different children responsible for observing and reporting about the weather at the *same time* each day and charting the information on a class weather calendar. Although the weather may change during the day (e.g., we have seen children want to add an afternoon thunderstorm to a sunny morning observation), it is important to track the weather at the same time. Discussing how the weather changes during the day can create another opportunity to explore weather.

As a class, decide upon a calendar legend. One approach is to use happy faces or sunshine symbols for sunny days and partially cover these based on the amount of clouds. The various symbols can be made blue to indicate a cool or cold day, while red or yellow could be used for warmer days; we used shading of sunshine symbols for temperature here. Snowflakes and raindrops can handle most precipitation events. For younger children, you should create the calendar the first month and then have the children follow your lead for the next. Older children can draw their symbols directly onto the calendar at the start of the activity.

Older children may also want to add thunderstorms, fog, sleet, and other weather variables (see page 15 for universal weather symbols). And they can use more than one symbol on any day. They can also discuss how the weather changed during the day.

Be sure to include the child's name on their reporting day (see sample for week of November 21 on calendar) to ensure "ownership" of the observations. The children really like this!

Each day, the child(ren) responsible for reporting on the day's weather should provide a class weather report. Their report should include words describing how the weather FELT to them (e.g., it was very cold and wet today; the wind blew leaves all around). If your school includes a weather report in its morning or afternoon "announcements," your weather reporter(s) could provide this service to the entire school.

Once a monthly calendar is completed, have children transfer the class data into a graph. Younger children can do this via one-to-one correspondence, and older children can count the number of days each particular type of weather occurred. The graph (page 11) uses the same symbols which appear on the calendar (page 10). However, a simple bar graph will also show the same results. Color-coding both the calendar data and the graph entries will make it easier for children the see the similarities and differences of displaying data in different ways.

For the graph on page 11, we keyed on the sky condition and weather; your children can further sub-divide these to classify weather days as either warm or cool.

Repeat the graphing activity after the second month's calendar is completed. Have children compare the graphs for both months.

After children have observed the weather for at least one week, have children predict what the weather will be the next day based on the previous data they have collected. Make a prediction graph showing the number of children that predicted the different types of weather. **Were their predictions accurate?**

Was the weather the same every day? If not, what changed? Which day's weather did you like best? What did you do on that day? Why did you like (not like) the weather on that day? How would you feel if the weather never changed? What if it was sunny or snowy every day? Was the weather the same or different during both months? In what ways?

Some of the answers to these questions can best be obtained by making live child graphs.

You can also do seasonal graphs with the children. Make seasonal paddles (paper plates glued or stapled onto tongue depressors or Kraft sticks. Have children draw seasonal pictures on their paddles. Have children line up in a human seasonal graph according to their favorite season.

> LEADER CHALLENGE: Save monthly weather graphs. In the spring, display graphs around your room with the month and year information covered. See if children can identify the month of each calendar by the weather information displayed on it.

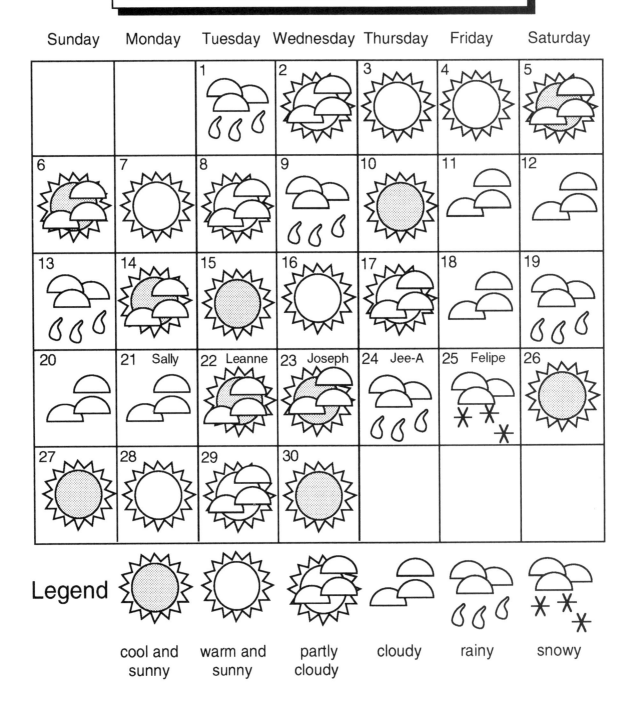

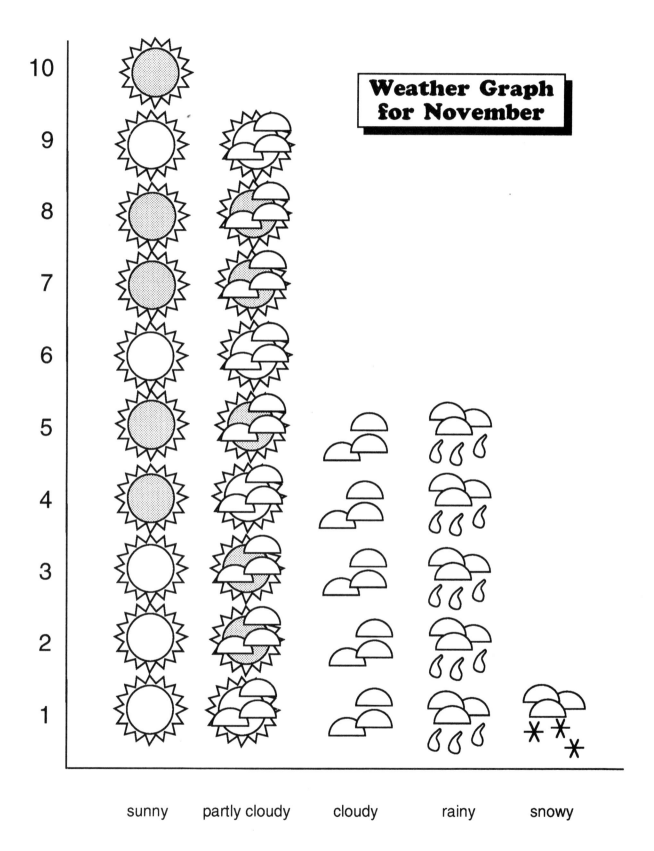

Weather Calendar for

Sunday	Monday	Tuesday	Wednesday	Thursday	Friday	Saturday

Legend

© 1996...H. Michael Mogil and Barbara G. Levine, 1522 Baylor Avenue, Rockville, MD 20850

4. Cloud Watching

Activity

Objectives
• to become aware of the different types of clouds
• to match and classify the cloud types

Materials • cloud flash cards (included), cloud chart, blanket

Post a laminated cloud chart or a set of cloud flash cards in your room. Encourage children to recognize the different types of clouds.

At a station or at group time, have children group the nine cloud flash cards which accompany this book. See answer key on next page. Once children have learned the three basic cloud families, you may also want to share the symbols/codes used by meteorologists to represent all the cloud types (see pages 14 and 15). Children may want to include this information on their daily weather calendar (Activity 3). You may also let the children group the flash cards by:

- geographic features (e.g., mountains, water, trees) • sky or cloud colors
- human factors (e.g., lamp posts, bridges, houses) • seasons
- any other criteria you wish

Weather permitting, go outside and have children lay down on a blanket and look up at the clouds. Bring out the cloud flash cards. Let each child have a turn trying to match the clouds that they see with those in the cloud pictures.

Were there any clouds in the sky? If so, were all the clouds the same type, the same shape, the same color? Were they all moving in the same direction? What do you think was making the clouds move? Could you find any clouds that had the shape of animals or things familiar to you? In order to make cloud classification easier for children, write down the words children say as they are watching the clouds. These words can be used to create simple cloud type descriptions and definitions. For example, children might use words like "puffy," "cotton-balls," "fluffy," or "soft" to describe cumulus-type clouds. These can be used in vocabulary building activities.

Would you like to be inside a cloud? What do you think it would be like to be inside one? If it's foggy, take the children outside to walk in a cloud! Have the children "feel" the moisture in the air, the moisture on their skin, and the moisture on various objects (e.g., leaves, playground equipment). **Do any of the children have a pet dog that went outside in the fog? Did the dog's fur get wet, too? What about the other animals that live outside?** Ask children if the car or bus they rode to school in had to use windshield wipers to clear away the water. **Where did the water came from?** Note that in fog, the moisture is NOT due to precipitation (Activity 8).

Suggested readings: *It Looked Like Spilt Milk*
The Cloud Book

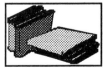 

CLOUD FLASH CARD ANSWER KEY

Cirrus -- High altitude ice crystal clouds (generally at least 15000 feet or 4500 meters above ground level). May appear as streamers, waves, curls, or chaotic masses. Often called "mare's tails," cirrus is a fair weather cloud unless it thickens or occurs with other clouds.

- Fort Worth, TX   #1
- Rockville, MD   #5
- Suitland, MD   #8

Cumulus -- Most often seen during the day, these cotton-puff looking clouds indicate rising air currents. Generally means fair weather, unless clouds become tall and dark. Cloud bases (generally 3000 to 6500 feet or 1000 to 2000 meters) are flat, showing the height at which condensation occurs.

- Poolesville, MD   #3
- Titusville, FL   #4
- Flagstaff, AZ   #7

Stratus -- A grayish, characterless sheet of low clouds (usually within 2000 feet or 600 meters of the ground). Only drizzle or snow grains will fall from true stratus clouds. When stratus clouds touch the ground, it is called fog.

- San Jose, CA   #2
- Washington, DC   #6
- San Francisco, CA   #9

Universal Cloud and Weather Symbols

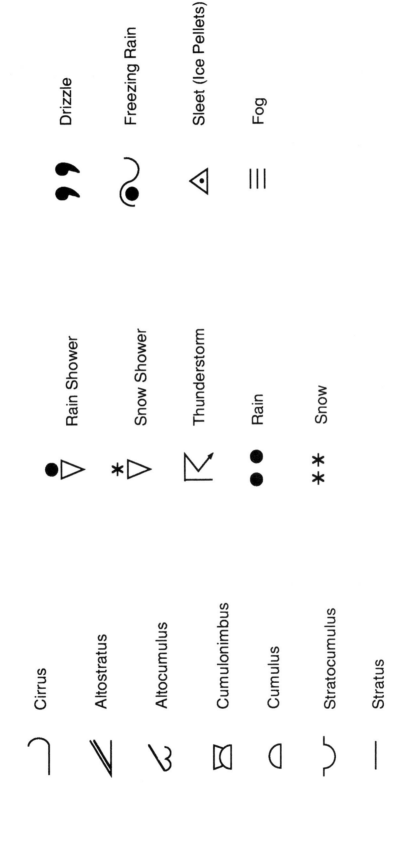

15

5. Creating Clouds

Activity

Objective • to create familiar cloud shapes

Materials • blue construction paper, white poster paint, small paper cups small mirrors, large container (e.g., roasting pan, sweater box), shaving cream, small cups, plastic spoons, blue Jello™, whipped topping

Activity #1 -- Give each child a piece of blue construction paper and a paper cup with a small amount of white paint. Have children fold their piece of paper in half the short way. Children should then open the fold and drizzle a small amount of paint onto one side of the fold. Be sure to keep the paint away from the edge of the paper. Fold and press down to blot the paint. Open and have children identify what they see. Label cloud pictures and allow to dry.

If the paint is drizzled along the fold, only one symmetric cloud shape will be formed; otherwise, mirror image pairs will be created.

After picture has dried, have children hold a small mirror vertically along the fold to recreate their symmetric shape or mirror image pairs. Then have them move the mirror around to create different shapes. **Will these be mirror images? Will these be symmetric?**

Discuss the cloud shapes the children make with what they have seen in the sky. Often cloud shapes resemble objects we are familiar with (see suggested reading).

LEADER CHALLENGE: What other objects can children find which contain symmetry in the classroom? outside? at home? This activity can be linked with Activities 1 or 2 or it can be done independently.

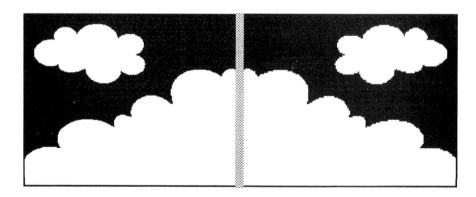

Activity #2 -- Fill a large container with water. Create cumulus clouds using small squirts of shaving cream. Allow small groups of children to blow the clouds around. Repeat as necessary.

Activity #3 -- Fill small cups with blue Jello™ and allow to gel. Serve to children. Add small amounts of whipped topping to simulate clouds. Enjoy the scenes; then eat them!

Activity #4 -- Fill small cups with blue Jello™ and add mini-marshmallows as it begins to gel. The "clouds" appear to float in the sky. As in the activity above, children can enjoy the beauty of their creations before they eat them!

Suggested reading: *It Looked Like Spilt Milk*

6. Creating Clouds (cotton balls)

Activity

Objective • to become aware of the different types of clouds

Materials • light blue construction paper, cotton balls, white glue, scissors, crayons, markers

Cut 12" x 18" construction paper in half lengthwise and prepare a tri-fold 6" x 18" construction paper for each child. Have children draw three scenes, one in each section of the paper. The first should be a cloudy day scene and the last two should be sunny day scenes. After drawing the scenes, have children glue cotton balls to represent the three basic cloud types. In the first scene, children should pull cotton balls apart and flatten them to represent layers of stratus clouds. In the second scene, children should simply glue the cotton balls in puffs to represent cumulus clouds. In the third scene, children should pull cotton puffs slightly apart to represent cirrus streamers and masses. Stratus means layers; cumulus means heaps or puffs; cirrus means wisps. Write the cloud name under each scene to reinforce the naming concepts.

Draw or let children draw shapes of animals or familiar objects on pieces of blue construction paper. Be sure to place name of the shape on the paper. Allow children to glue cotton balls onto their shapes to create clouds. Often cloud shapes resemble objects we are familiar with (see suggested reading).

Suggested readings: *The Cloud Book* (note - for younger children, we recommend skipping pages 12 -15 in this book because the material is too complicated)
It Looked Like Spilt Milk

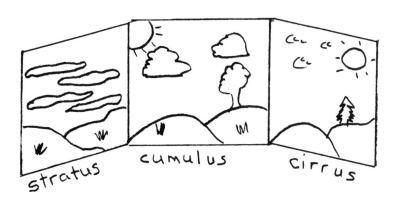

7. Creating Clouds (in a bottle)

Demonstration

Objectives
• to observe how a cloud forms
• to observe cloud droplets in motion

Materials
• clear 2-liter plastic soda bottle, very warm tap water, flashlight, matches or chalk

Discuss with children the recipe for creating a cloud. Simple definitions of water vapor (water in a gaseous form) and condensation nuclei (e.g., small, solid objects in the air, like dust, pollen, volcanic ash, salt from the ocean spray, and smoke) will emerge. To demonstrate water vapor, talk about steam rising from boiling water or from the street after a summer shower. To demonstrate condensation nuclei, watch sunlight shining through a window or watch the sun's rays shining among the trees. Look for smoke stacks. Clap two erasers together. Gather dust or pollen from a surface in your classroom or outside. If you live near the beach, observe what happens to eyeglasses and car windows.

The most difficult ingredient might involve temperature. Because they see steam, they may think that warm temperatures are needed to make clouds. Actually, cool air is needed. Think about how condensation occurs on the outside of a cold can, bottle, or glass on a warm, humid summer's day.

Then tell children that you can create a cloud in the classroom. To do this, add about an inch or two of very warm tap water to a 2-liter plastic soda bottle. Quickly cap the bottle. Gently squeeze and release the bottle. The squeeze simulates "warming" that occurs in the atmosphere; the release, the "cooling." Observe what happens. If the inside of the bottle becomes covered in condensation, just shake the bottle and repeat the squeezing and releasing. A cloud will NOT appear at this stage. This is because you have added water vapor to the air, but you have not yet added anything onto which the water vapor can condense (other than the bottle). The children will be excited that your demonstration doesn't work! Then, act surprised when you realize that you left out an ingredient.

| cloud | tornado | blizzard |

- water vapor
- condensation nuclei
- cooling of the air

Open the bottle. Carefully light a match and hold the match near the mouth of the bottle. Blow out the flame and blow some of the smoke into the bottle. Then drop the match into the bottle (*for safety*) and quickly close the cap. (Note: chalk dust works, too). Now gently squeeze and release the bottle again. The children should see a cloud appear (release) and disappear (squeeze).

For older children, pass the bottle around and let each child have a turn. For younger children, place the bottle on an overhead projector (with the light on), or shine a flashlight at the bottle to better see the cloud and its movement. You can also hold the bottle against a dark background. Clouds are composed of small cloud droplets like these, and may also contain raindrops, snowflakes, ice pellets and/or hailstones.

This demonstration shows that water vapor (water in its invisible gaseous state) can be made to reappear as small cloud droplets. The process can be speeded up by adding small condensation nuclei (particles, like dust or smoke) onto which the water vapor condenses.

What was the cloud in the bottle made of? How did the cloud droplets move? Was it easier or harder to see the cloud droplets when smoke was added to the bottle? What can you see floating in the air when the sun's rays shine through a window? Would the cloud be easier to see if two or three matches were used?

LEADER CHALLENGE: Have children look for clouds in their homes. Hint: the kitchen and bathroom are good places to investigate.

Have children make their own clouds outside. On a cold morning or on a chilly, rainy day, have them go outside and "puff". Discuss how this cloud forms.

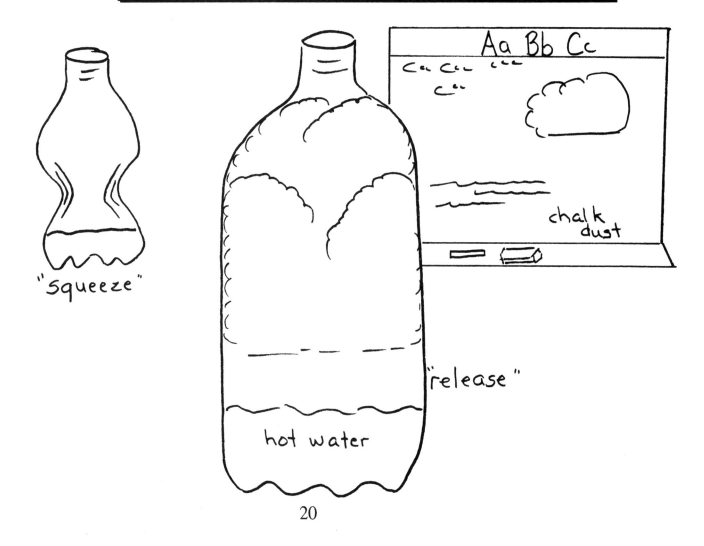

8. I'm In a Fog

> **Activity**

Objective • to become aware that fog is one type of cloud

Materials • light blue construction paper, crayons, markers, wax paper, white glue

Brainstorm with children what it might be like inside a cloud. Clouds are composed of millions of small water droplets (sort of like drizzle drops). These float around due to wind currents and don't fall out of the clouds unless they become larger and heavier. On a foggy day, you can't see as far because cloud droplets get in the way. The more cloud droplets, the denser the fog and the longer it will usually last.

Make a foggy day picture. Color a sunny day landscape scene on a piece of light blue construction paper. Cover half the picture with a piece of wax paper by gluing one edge of the wax paper onto the edge of the construction paper. Compare the scene with and without the wax paper. The wax paper acts like fog.

Suggested reading : *Fog* (a poem by Carl Sandburg)

> The fog comes on little cat feet.
> It sits looking over harbor
>   and city on silent haunches
>   and then moves on.

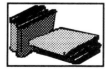 

9. A Cloud and Sky Watcher Window

Activity

Objective • to become aware of the different colors of the sky and clouds

Materials • poster board, poster paint (white, blue and black), paper plates, paint brushes, crayons, markers, paint chips, glue sticks

Prepare sky and cloud watchers using 8 1/2" x 11" poster board. Fold poster board in half; along the folded edge, cut out a rectangle leaving about a 2" frame.

One way to make a cloud and sky watcher window is to paint half the watcher in blues and the other in grays. Alternatively, paint one side of the watcher in blues and the other side in grays. Be sure to leave one area unpainted to represent white. To do this, for each child, place 6 to 12 teaspoon-size blobs of white poster paint around the outside of a paper plate, and a larger blob of blue paint in the center. They should then add one drop of blue paint to the first blob and gently mix. Then they should paint a large band (about 1" to 2") from the outer edge to the inner edge of the watcher. They should then add two drops of blue paint to the second blob, and paint another band next to the first. Continue painting successively darker shades until all white blobs are used up.

Repeat the activity using black paint.

Allow to dry and number or letter the colors in order so that the children can easily identify the colors they see.

You can also use paint chips. Make a collection of blue and gray paint chip strips from your local paint store; be sure to include at least one white. Give each child a varied selection of blues and grays. Have them cut the colors apart, leaving the color names on the chips for easier color identification. Using a glue stick, they should paste the chips onto their sky and cloud watcher window frame.

Have children predict the color(s) they expect to see most frequently. Post their predictions.

Go outside and have children hold their watcher windows up to the sky in the direction *away from* the sun. Have them match (as best they can) the actual sky or cloud colors with those on their watchers. Tell the children that they are each watching their own personal piece of the sky. It is okay to find more than one color in the sky or in the clouds!

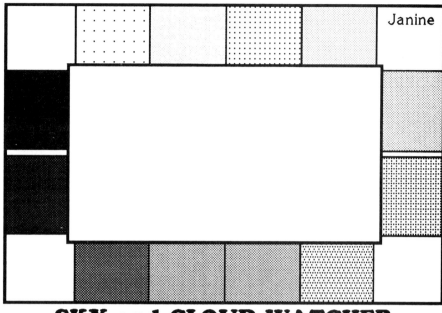

SKY and CLOUD WATCHER
This sample watcher window was created using computerized shading

Make a master sky watcher window for your class. As a class activity, assign a daily weather observer (or team) to record observed cloud and sky colors on the daily weather calendar. Children can often see more than one color during their observations. The data can later be transferred to a separate data record (see below) for easier counting / charting at the end of the month. **How did the observed sky and cloud colors compare to their predictions?**

Children can also compare weather to sky and cloud colors. For example, **under what weather conditions is the sky the brightest blue? the darkest gray?**

You can extend this activity to "colors in nature" (Activity 2). Create nature watcher windows for your particular locale and season. Use reds, yellows, greens, and browns for fall colors in many parts of the country. For desert areas, we suggest browns, yellows, and muted reds and oranges. For snowy regions, make a watcher with different shades of white.

Does everyone see the same colors in the sky? Why or why not? What might be causing the different colors (e.g., pollution or different amounts of sunlight)? Remember that each child looks at a different part of the sky and the ability to distinguish colors is different for each person. Each child really sees the sky and clouds as no one else can!

> LEADER CHALLENGE: Have children predict the color change from one day to the next based on at least one weeks worth of observed data. **How accurate were their predictions?**

SKY and CLOUD WATCHER DATA RECORD

date	blue haze	sky blue	azure	bluebonnet	misty	rainy day	charcoal	milky
Monday Oct 20						X	X	
Tuesday Oct 21				X				
Wednesday Oct 22				X			X	X
Thursday Oct 23				X				X
Friday Oct 24		X						
Monday Oct 27						X	X	
Tuesday Oct 28		X						
Wednesday Oct 29		X			X			
Thursday Oct 30		X			X			X
Friday Oct 31	X							
count	1	4	0	3	2	2	3	3

SKY and CLOUD WATCHER DATA RECORD

date

count									

© 1996...H. Michael Mogil and Barbara G. Levine, 1522 Baylor Avenue, Rockville, MD 20850

10. Sky Diary

Activity

Objectives
- to record cloud and sky observations in written or art form
- to make weather predictions

Materials
- white construction paper, crayons, markers, lined paper (optional), stapler

Prepare a cumulus-shaped sky diary cover using white construction paper. Cut out at least five pieces of lined paper in the same shape for each book. Staple together. Children should name their book (e.g., "John's Cloud Book" or "Maria's Sky Diary").

Each day, have children write or draw a description of the sky. They should write the day of the week and the date on each page and write about how the sky made them feel.

As in Activity 3, if you use this activity for more than one week, have children predict what the weather will be the next day based on their previous observations.

See Copycat Magazine, March/April 1995 for other sky diary ideas.

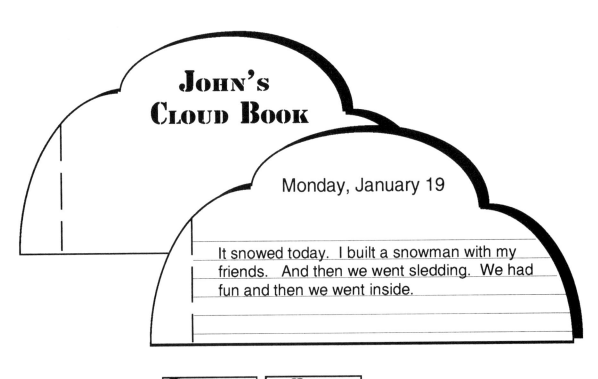

11. Raining Cotton Balls

Experiment

Objectives
- to predict how many cloud droplets it takes to make a cotton ball cloud rain
- to collect and compare data
- to explore how "variables" can affect the outcome of an experiment

Materials
- cotton balls, pipettes or eye droppers, paper cups, water, wax paper, data sheet

Familiarize children with pipettes or eye droppers. Ask them if they have seen a similar device around their kitchen (e.g. turkey baster). Show them how to use the pipette by providing each child a small cup of water and a pipette. When squeezed in the water, air is pushed out; when released in the water, water is pulled in; outside the water, the reverse occurs. Let the children practice using their pipettes. They should be able to control the pipette so they squeeze one small drop at a time.

Discuss what clouds are (masses of cloud droplets) and that clouds only rain when there are too many drops in them.

Pair children. Give each pair a cotton ball "cloud" and a piece of wax paper. Give each child their own data sheet. Have children predict how many drops from a pipette it will take to make their cotton ball "cloud" rain. Children should record their predictions. Have one hold the "cloud" over the wax paper and the other squeeze drops of water onto the "cloud." The child adding the drops should count the number until the other child announces that the cloud "is raining." Children should write down the actual number on their data sheet. Note that the data sheet has been designed to be used for a single class experiment or for children repeating an experiment to study variables.

Give each pair a new cotton ball. Have the children again predict the number of drops it will take to make the new cloud "rain." Repeat the experiment having the children reverse jobs. **For which experiment were the children's predictions most accurate?** Note that the children had more information on which to base their predictions in the second experiment!

Have each pair take their wet cotton balls and squeeze out as much water as they can. Have them predict how many drops will be needed for these used cotton ball "clouds" to "rain." Repeat the experiment.

Ask children to "adjust" their cloud (e.g., stretching it out to make a stratus cloud). Have them predict whether a cotton puff (cumulus) or a stretched out (stratus) cloud will hold more water. Have the children design an experiment to verify this.

How close were the predictions to the actual observations? Did each cotton ball hold the same number of drops? Compare several dry cotton balls - are they about the same size? Have several children each squeeze a single drop of water onto a piece of wax paper. Examine the drops; **are they about the same size?** Two variables which can affect the outcome of this experiment are the size of the cotton balls and the amount of water in each pipette squeeze.

> LEADER CHALLENGE: Brainstorm with children about other variables that could have affected the outcome of their experiment. Answers could include where the cotton balls were held (edge, center), the orientation of the cotton balls (horizontal or vertical), and the distance between the dropper and the cotton ball. Have children design other experiments, which help to control some of these variables; see if the results change. Test cotton balls and cosmetic puffs to see which hold more water.

"RAINING COTTON BALLS" PREDICTION DATA SHEET

	dry cotton ball			
name	Yolanda	Kevin	Tai	Christie
predicted	34	10	27	63
observed	23	35	21	30
difference	11	25	6	37
	squeezed cotton ball			
name	Yolanda	Kevin	Tai	Christie
predicted	20	25	17	25
observed	21	25	18	20
difference	9	0	1	5

Sample class data sheet: experiment using dry cotton ball (top) and reusing it after squeezing out water (bottom)

_____   _____
name   date

"RAINING COTTON BALLS" PREDICTION DATA SHEET

name				
predicted				
observed				
difference				
name				
predicted				
observed				
difference				

Method Used: _____
_____
_____
_____

Observations: _____
_____
_____
_____

Conclusions: _____
_____
_____
_____

© 1996...H. Michael Mogil and Barbara G. Levine, 1522 Baylor Avenue, Rockville, MD 20850

12. Keeping Raindrops Up!

Demonstration

Objective • to examine one way that raindrops stay in clouds

Materials • hair dryer (preferably with burned out heating element or a "cool" button), 2 ping-pong balls, tennis and/or golf ball

Discuss wind and the fact that it can blow up and down, as well as horizontally (Activity 36). Aim a hairdryer straight up and display a ping-pong ball. Ask children what they think will happen when the hairdryer is turned on low speed and the ping-pong ball is placed in the current of rising air (called an updraft). After showing what happens, ask children what they think will happen if the hairdryer is turned on high speed and the experiment is repeated. Show what happens.

Repeat the experiment again using the high speed setting. This time gently tip the hairdryer so the wind doesn't point directly upward. Discuss what happened. Repeat again. With the hairdryer pointing directly upward, add a second ping-pong ball. Finally, repeat the experiment using a tennis or golf ball to simulate a large raindrop or hailstone.

Raindrops (and snowflakes and hailstones) stay in the cloud until they become so heavy that gravity overcomes the vertical wind. Since thunderstorms have strong updrafts, large raindrops and hailstones often accompany them.

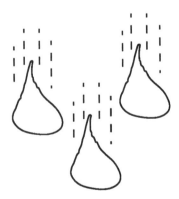   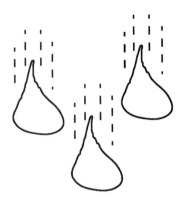

13. The Great Raindrop - Snowflake Race!

Activity

Objective • to examine how raindrops and snowflakes fall

Materials • pieces of paper, scissors, magnifiers, tongue depressors, black construction paper, white glue

Brainstorm several characteristics of both rain and snow (e.g., color, shape, when they occur). Make a list of these and post in your classroom.

Discuss how rain and snow fall. Ask the children which they think falls faster. Make a human graph and/or chart their predictions. Pair children, giving each pair two identical pieces of paper. Crumple one, and leave the other flat. Ask the children which of these would represent rain and which snow. Have each pair hold their two pieces of paper at the same height above the ground and then release them simultaneously. **What do the children observe? Which fell fastest?** Graph results.

Since all of the pieces of paper are the same weight (or if they are different, it should be a random difference), first indications are that they would fall at the same speed. **What other variables can be causing the difference?**

The crumpled piece of paper has less frictional drag (or air resistance) on it than the flat piece of paper. Snowflakes tend to be flat, and even in clusters, they are flatter than more aerodynamically-shaped raindrops. Hence, raindrops will fall faster than snowflakes.

If there is day when rain and snow are falling together, have children watch. **What do they observe?** If they see snowflakes winning the race, it's probably because the larger clusters of sticky snowflakes are heavier than the raindrops.

If children want to change a variable, have them cut a piece of paper into a snowflake design. Cut off part of the flat piece of paper to equalize the weight. Then hold a race between the flat piece of paper and the snowflake. **Which do the children think will fall fastest?** Graph predictions and results. Discuss outcomes.

Take advantage of snowy days to enhance your children's observational skills. Go outside and have children catch snowflakes on their jackets, mittens, and hats. Being careful not to breathe on the snowflakes, have children examine them using their eyes alone and magnifiers. Alternatively, have them build snow catchers using black construction paper (about one or two inches square) attached with glue to the end of a tongue depressor. **What do they notice? Are all snowflakes six-sided? Are there any that look alike?**

Examine clumps of snowflakes too. **How do the snowflakes stick together?** Simulate this process by having children cut out snowflake designs and allow the points to intertwine. Dampen some of the edges of cut out snowflakes and try to put them together with other dampened flakes. **Do they stick?**

Water

14. Sky Inventory

Activity

Objectives
• to heighten awareness of things found in the sky
• to assess prior knowledge

Materials
• chart paper, markers, crayons, small pieces of paper, index cards

Ask children what can be found in the sky. Write down all responses on one large sheet of chart paper and post it. You can also use an overhead projector. Have each child copy the responses onto individual pieces of paper or index cards. Then ask the children to group and classify their responses. Have children report on their classifications and the reasons for them. You will find that there will be many valid classifications. Show these new groupings on a different sheet(s) of paper. If children have difficulty writing, limit the number of responses to about twelve. For younger children, you may want to use pictures along with the words to help group the responses (including grouping them as they are given).

Some groupings could include "astronomy," "things that are alive," "things that fly," and "machines." Some responses may fit into more than one group (an opportunity to use Venn diagrams). Be open for unexpected or unusual responses.

The most likely grouping that will emerge is "things that belong to the water cycle." Evaporation, condensation, precipitation, rain, snow, pollution, dust, clouds and the sun will be among the words you hear. Thus, this activity becomes the natural lead-in for the "water cycle" activities that follow. If the "water cycle" does not materialize, create it yourself.

15. Water Cycle Web

Activity

Objective • to explore how things in the water cycle are related

Materials • index cards, markers, scissors, yarn or string

Ask children what things are "in the sky." Write the word, draw a picture, and place the child's name on a large index card for each response for younger children; older children can complete this themselves. Be sure that there are no duplicate responses. After all responses are given, have children sit in a large circle. Select the child with the "sun" to place their card on the floor first. If a child did not select the sun, you should place a sun card on the floor. Ask children **"who has something that is related to the sun?"** Choose one child and have them place their card next to the sun. Connect the two cards with a piece of yarn or string. Then ask **"who has a card that is related to either the sun or the card that was just placed on the floor?"** Continue to connect the cards until all children have placed theirs into the web. Discuss the web. Ask if anyone wants to move their card to another place. Before they do this, ask them how it might affect the web. If it doesn't "mess things up," let the child move the card. You'll find an infinite number of potential groupings emerge, including the "water cycle.

See Activity 52 for a complementary solar web activity.

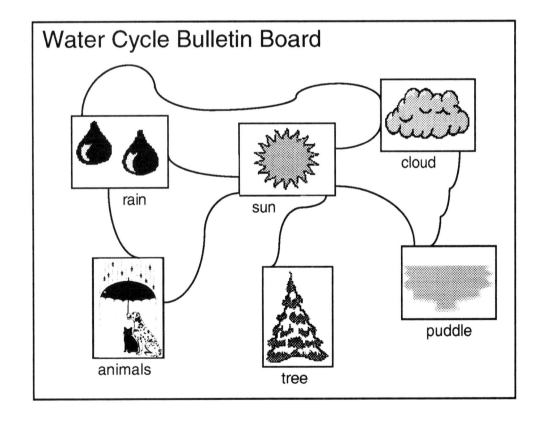

16. Around and Around the Water Cycle

Discussion

Objective • to understand the water cycle

Materials • large white paper, markers, crayons, chalk board, chalk

Ask children where they have seen rain or snow. For example, they may have seen rain or snow falling, splashing in puddles, or running down the street. They may have also heard the sound of rain hitting the roof or windows of their home. This begins following the water cycle through its four steps - precipitation, accumulation (or collecting), evaporation, and condensation (see Activity 14). Whatever the initial response, you can continue the cycle from that point. Based on this discussion, draw the water cycle (using many colors) until you have included all components. Remember that a heat source (e.g., the sun) is needed for evaporation to occur. This should also appear on your drawing. Although it won't be used in the discussion at this time, note that a cold source (e.g., cold ground or cold air in the upper parts of the atmosphere) is needed for condensation to occur.

Ask children where they think water accumulates. Typical answers include puddles, rivers, oceans, and plants. Don't forget ground storage, glaciers, and even people!

To emphasize the "cycle" part of the water cycle, show that the four steps are joined and repeat themselves over and over. You might even want to tell the children that they may be drinking the same fresh water today that dinosaurs drank millions of years ago!

> LEADER CHALLENGE: Have children trace the movement of water through their local area, including typical water cycle processes. What happens to water they use in their homes and at school, and how is water "treated" and then added back into the natural water cycle? You can also take a field trip to a waste water treatment plant.

17. It's Raining Without Raining

| Activity |

Objective • to use creative movement to simulate sounds associated with precipitation

Materials • none

Have children sit in a circle on the floor. Tell them they are going to make a rainstorm, but that they have to listen very carefully. The steps can be done either in sequence as a group or in a round. The children will first make the sound of wind rushing through the trees, then the sound of individual raindrops followed by heavier and heavier rain, and finally the sound of thunder (see below for suggestion for adding lightning). Then reverse the sounds to simulate the storm moving away from the children.

The sequence of sound-making activities is:
- Rub hands together gently
- Snap fingers or tap two fingers onto a hard surface (e.g., desk or palm of other hand)
- Slap hands on thighs
- Slap hands on thighs faster and stamp feet
- optional - turn light switch on and off rapidly for lightning effect; continue slapping and stamping
- Slap hands on thighs
- Snap fingers or tap two fingers onto a hard surface
- Rub hands together gently
- Stop

And nobody even got wet!!!

We have found that children like to do this several times. Repeat as needed. Do the activity at least twice BEFORE trying it as a round.

18. Now It's Really Raining!

Demonstration

Objectives
- to make it rain indoors
- to visualize the various parts of the water cycle

Materials
- electric skillet or portable electric burner and a small pot, water, ice cubes, oven mitt, small frying pan

Heat water to boiling in either electric fry pan or in the small pot. (Hint: begin to heat water several minutes before gathering children). With children sitting at a distance, show them the vapor coming from the water. This simulates sunlight shining on a body of water, causing the water to evaporate. The steam that comes off the heated water is visible for only a second or two before it becomes invisible. Tell children that you can make this water reappear!

Show children an empty frying pan. Ask a volunteer to examine the pan and look for any holes. You can also knock on the pan (inside and outside) to prove that it is solid. To further verify that there are no holes, you can add some tepid water to the pan and have another volunteer check for water leaks. Then pour off water.

Add ice cubes to the frying pan. Put on your oven mitt and hold the pan a few inches above the boiling water. As the warm, moist air rises, it is cooled by the chilled frying pan. This simulates the condensation process which occurs in the atmosphere. After a few moments, lift the pan up and have children observe the bottom of the pan. **What do they see?** (Small droplets of condensed water should have formed). Return the frying pan to its position above the boiling water for a few more moments. Again, look at the droplets (which should be bigger this time). Repeat until droplets become large enough to fall as raindrops. Some children may want these drops to fall on them. You can also let the drops fall back into the boiling water, where they are then re-evaporated. This simulates the entire water cycle.

What was on the bottom of the frying pan? How did it get there (especially if there weren't any leaks in the frying pan)? How did the little droplets become big enough to fall as raindrops?

19. A Mushroom in the Rain

Activity

Objective
- to show the connection between language arts, drama and science
- to demonstrate one effect of rain on plants

Materials
- parachute, the book *Mushroom In The Rain*

Read the book *Mushroom in The Rain*. Ask if any children have ever seen mushrooms growing. **Were they bigger before or after the rain?** (*NOTE - be sure to mention to children that they should NOT eat mushrooms they may find. After all, the animals did not eat the mushroom in the book..*)

To test the hypothesis, bring in some real mushrooms (store bought). Soak a few in water. Let children use magnifiers to observe differences between the soaked and the unsoaked mushrooms.

Produce the story as a play for the class. Assign the parts of the animals. The rest of the class will serve as the holder of the mushroom. Repeat the play several times so that everyone has a chance to be an animal.

Use a parachute as the mushroom cap. Roll the parachute under at the edges, so that it can be unrolled as it "rains." The audience will unroll the mushroom as the story unfolds.

Read the story as a narrator and have the children walk on "stage" and get under the parachute mushroom as their character is mentioned; of course, the fox does not get under the mushroom. Older children can say their lines (provided on index cards). You can enhance this activity by using animal masks (made from paper plates), simple animal features (e.g., antennae, wings, ears, tails, and noses), or simple animal costumes.

Have children take turns being the narrator and retell their version of the story. How would they change it?

Suggested reading: *Mushroom In The Rain*

20. Playing with Raindrops

> **Activity**

Objectives
• to see how cloud drops become raindrops
• to learn about some properties of water and other liquids

Materials
• misters, water, wax paper, coffee stirrers, straws or pipettes, other liquids (e.g., cooking oil, lemon juice, vinegar)

Fill misters nearly full with water and adjust sprayer so a fine mist comes out. Tear or cut a wax paper sheet for each child (approximately 12" x 12"). Give each child a coffee stirrer, straw, and/or pipette. Spray water three to five times onto the wax paper to create cloud droplets. Have children observe their drops. Make an ongoing list of observations about the drops (e.g., round, have air bubbles in them, clear, are different sizes). Ask children to play with the drops. You will find that some children blow the droplets together. Others may lay the coffee stirrer down and plow through the water to create the biggest possible drop! Others may use the pipette as a grabber to pull the water along or to pull it up. You may want to discuss these various techniques in a group setting.

Regardless, the outcome is similar - the cloud drops become bigger (i.e. they become raindrops).

What happened when a cloud drop came close to another cloud drop? Why did this happen? (surface tension - the skin on one water drop attracts the skin of another. That's why big water drops don't fall apart).
Have children repeat the experiment using different liquids. You should drop these liquids onto different pieces of the wax paper. **Do the drops behave the same way as the water drops? Did they move around as easily?**

> LEADER CHALLENGE: Tell children that their drops can be used as magnifiers. Ask them to predict which type of drop (large or small) they think will be a better magnifier. One option is to take a poll and post results. Then, cut up and distribute parts of the classified advertisements from the newspaper. Have children slide their piece of newspaper underneath their wax paper. Instruct them to recognize letters or read words through large and small drops. **What do they notice?** They should recognize that the smaller drop is a better magnifier! To understand why, have children observe drops from table level. They should see that the smaller drop is rounder than the larger one. If any children wear glasses (or if you do), have children examine the shape of the lenses. **What else could the children observe under water drops?** (e.g., coins, comics)

21. Surface Tension

Experiment

Objectives
- to explore surface tension
- to learn more about properties of water
- to explore how "variables" can affect the outcome of an experiment

Materials
- coins, pipettes or eye droppers, small paper or plastic cups, lots of paper clips, milk, Pyrex™ pie plate, dishwashing detergent, food coloring (4 colors)

Experiment #1 -- Give each child a penny, a pipette, and a cup of water. Allow children to explore how to use the pipettes. They should be able to control the drops.

Give each child a data sheet (see Activity 11). Have each child predict how many drops of water they can get to fit on the head of their penny before the water falls off. Due to surface tension, the drop will become quite large and actually overhang the edge of the penny before the surface tension breaks.

Repeat the experiment using pennies from different years. Try the experiment using a penny from the year the child was born. Try the experiment using very old and very new pennies. You may also want to try different coins,

"SURFACE TENSION" PREDICTION DATA SHEET

1985 pennies				
name	Ginger	Ernesto	Alex	Janis
predicted	16	2	6	22
observed	23	25	29	27
difference	7	23	23	5
1995 pennies				
name	Ginger	Ernesto	Alex	Janis
predicted	24	25	27	29
observed	30	29	28	31
difference	6	4	1	2

Sample class data sheet: experiment using pennies from various years

including coins from other countries. Discuss results. As with Activity 11, this activity demonstrates how "variables" can affect the result of an experiment.

Use a salt-vinegar solution to clean the pennies. Then repeat the experiment. **What is different? How, if at all, did this change the outcome?**

Experiment #2 -- Give each child a cup of water filled to the lip, a large number of paper clips, and a data sheet (see experiment #1 above). Have each child predict how many paper clips they can add to the cup of water before the water overflows. Paper clips should be added slowly at the edge of the cup, one at a time. As with the pennies, have children observe how the water builds up and overhangs the edge of the lip of the cup.

Demonstration -- Gather children. Place a thin layer of milk into a Pyrex™ pie plate. Add one drop of red food coloring at the 12 o'clock position. Add drops of blue, green, and yellow to the 3, 6, and 9 o'clock positions. By the way, there is nothing magic about this order! Ask children what they observe.

Have children predict what they think will happen when you add a small drop of dishwashing detergent (we prefer Dawn™) to the center of the pie plate.

Add the detergent and observe. The reason the patterns develop is that the detergent breaks down the surface tension of the milk. This is why we use soap for washing dishes and for washing ourselves.

Activity -- Make a soapy solution using Dawn™ or another dishwashing liquid. Have a child use a Q-Tip™, dipped in the solution, write a message on the bottom of a frying pan. Then use the apparatus from Activity 18 (without ice cubes) to create a small amount of condensation on the bottom of the frying pan. Tilt the pan to read the message. **Why didn't the water condense into small droplets where the writing was?**

_____ _____
         name   date

"SURFACE TENSION"
PREDICTION DATA SHEET

name				
predicted				
observed				
difference				
name				
predicted				
observed				
difference				

Method Used: _____

_____

_____

_____

Observations: _____

_____

_____

_____

Conclusions: _____

_____

_____

_____

© 1996...H. Michael Mogil and Barbara G. Levine, 1522 Baylor Avenue, Rockville, MD 20850

22. Fossilized Raindrops (A great activity for outdoors)

Activity

Objectives
- to examine individual raindrops
- to preserve raindrops
- to compare different rain events

Materials
- aluminum pie tins, corn starch, strainers, misters, water, small zip-lock sandwich bags, permanent marking pen or labels

Prior to beginning this activity label small zip-lock sandwich bags with children's names and the words "fossilized raindrops".

Divide the children into small groups and provide each group with two aluminum pie tins and a mister filled with water. Adjust misters so they spray a fine to medium spray. For each group, strain a thin layer of corn starch into one of the pie tins. Have one child be a rain maker (give that child the mister) and another be a rain collector (give that child the pie tin with the corn starch). The other children in the group should be observers. Each rain maker should aim their mister toward the ceiling (or if outside, toward the sky) and spray five times. The rain collector should try to catch as many raindrops as he/she can.

Have children observe the pie tin used to collect the raindrops. **What do they notice?** Some children may see the raindrops not yet coated in corn starch. Others may see "impact" craters, similar to those found on the moon or when rain falls on sand or clay soil.

Assist each group in straining the corn starch from their pie tin into the other pie tin by pouring the corn starch through a strainer; gently shake and tap the strainer to get rid of excess corn starch. What will remain will be raindrops coated (encased) in corn starch. Carefully pour these into the rain collector's sandwich bag. Refill pans with corn starch and have children rotate to a new position (e.g., the rain collector now becomes the rain maker, the rain maker becomes an observer, and one of the observers becomes the rain collector). Repeat until all children have collected their own "fossilized raindrops".

Have children observe their "fossilized raindrops." Discuss with children why these are "fossilized." (Fossilized means preserved either directly or via an imprint. These drops leave their imprint inside a corn starch covering. If left undisturbed, these fossilized drops could be preserved for years). Create your own bag of "fossilized raindrops" and display it on a class bulletin board for later use.

Are all raindrops in a given bag the same size? Are raindrops in all bags the same size? How many drops did each child collect? Did each child collect the same number of drops? How many drops do you think fall when it rains? (remember that the drops collected here came from only five sprays from a mister). **Were the variables in each groups experiment identical? If not, what could have been different?** Graph results.

> LEADER CHALLENGE: If it rains or snows, repeat the activity by collecting real raindrops or snowflakes. Have children stand outside for a count of five. Have children compare these samples with what they collected outdoors on another day or with what was collected during the classroom activity. Have children compare and contrast snowflakes* and raindrops. Note that children will not collect standard looking snowflakes. This is because the snowflakes must start to melt first for the corn starch to stick to them. Also, snowflakes often fall in clumps rather than individual flakes.

DO NOT perform this activity in a thunderstorm!

* To collect snowflakes, you will need to chill the pin tins and corn starch by leaving them outdoors in a protected location for at least an hour before using them.

23. Measuring the Rain

Activity

Objective • to measure how much rain falls

Materials • clear, plastic 2 liter soda bottles, rulers, water, index cards, data sheets

Collect enough clear, plastic 2 liter soda bottles for all children in your class. Carefully using an X-ACTO™ knife or a single edge razor blade, cut off the top third of the bottle. Be sure that you cut just below the slight ridge present just beneath the sloped bottle top. Invert the top piece so that it rests snugly in the bottom part of the bottle. This creates a rain gauge; the inverted cover helps lessen the amount of evaporation.

Put the covers away and fill the gauges with different amounts of water. Place these around your room, each on a lettered index card. Discuss what a rain gauge is and why it is important to know how much rain falls. (A rain gauge is simply a collector of rainfall. Knowing the amount of rain that falls is important to farmers, meteorologists, dam or reservoir operators, and boaters). Show children how to use a ruler to measure rainfall.

Give each child a ruler and a data sheet (see page 41). Have children measure the "rain" in each gauge and record their measurement. They should also shade in the picture to represent the amount of water in that gauge. This will give them practice in reading a ruler and in measuring rainfall.

Brainstorm how this rain gauge could be used to collect real rainfall. Discuss where children could place their gauge to accurately collect rainfall (e.g., away from trees, not under the eaves of a building). If it rains, set a container outside and collect and measure rainfall. Because the bottom of a soda bottle is not exactly flat, add several inches of water (or enough water to show above the colored base of some soda bottles) to control your measurements and to weigh down the bottle. Be sure to subtract this amount of water from the total in the container to obtain the actual rainfall. This data could be added to the daily weather chart (Activity 3) or it could be graphed or charted separately.

> LEADER CHALLENGE: Have children take their rain gauges home. For a period of several weeks, they can measure the rain with the help of their parents and share observations in class. Be sure to remind them to partially fill their gauge first and to subtract the initial water depth from their final measurement! **Did the same amount of rain fall at each child's home? Did it rain at each child's home? How did the measurements of the children compare with the official National Weather Service observation (usually taken at the airport)?** This information may be found in the local newspaper or on television weather reports.

name _____ date _____

Measuring Precipitation

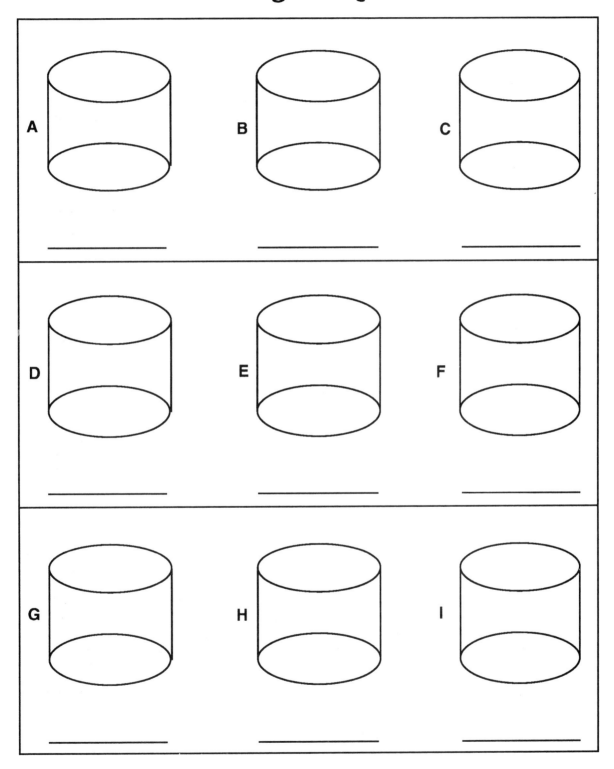

© 1996...H. Michael Mogil and Barbara G. Levine, 1522 Baylor Avenue, Rockville, MD 20850

24. How Much Water Is In My Snow?

Experiment

Objective • to predict how much water is in snow

Materials • clear cylindrical plastic container, rulers, hair dryer, snow or crushed ice

Fill a container (a two to four inch high plastic specimen container works well) with packed snow or crushed ice so that the contents extend slightly above the top of the container. Show this to the children and ask them what they think will happen when the snow or ice melts. Have them predict how much water will be left afterwards. For younger children, you can give three choices (e.g., the container will be full of water, half full of water, and the container will contain only a small amount of water) or allow them to predict freely. Ask if anyone thinks the container will overflow. Post a record of their predictions. You can adapt the data sheet from Activity 23.

Carefully use a hair dryer to melt the snow/ice or place the container near a radiator, heating vent, or sunny window. As the snow/ice melts children will likely comment about what they are seeing. Some may want to change their predictions. This is okay, but do not change the posted prediction record.

An alternative is to let the snow melt naturally and have children mark the height of the water on a data sheet. Check every half hour.

Have one child measure the depth of water remaining after the snow/ice melted. For example, if your container was 10" inches tall and filled with packed ice, the ice should melt to about 10" of water (making the container full). On the average, about 10" of snow will melt to 1" of water.

> LEADER CHALLENGE: Place container with the melted water in a freezer (or outside in a sheltered location if the temperature is cold enough) and refreeze the water. **Did it fill up the container?** Discuss what might have been trapped in the packed snow and ice that isn't in the water (air). Children can also see the effect of trapped air when they walk on snow or when they make snowballs (the snow compacts).
>
> Bring in two pints of ice milk manufactured by different companies. Place each on a pie tin and allow to melt. Have children predict whether the ice milk will be warm or cold when it melts. Pour the liquid from each into the same sized container. Compare the amount of liquid in each. **Were they the same? different? Which brand of ice milk had more ice milk? Which had more air bubbles?** Pour into individual paper cups and serve as a snack. Be sure you have enough for everyone in the class.

Suggested reading: *The Snowy Day*

After reading *The Snowy Day* have children discuss ways that they can keep a snowball for another day. If children suggest putting it in a freezer, see Activity 29 for related activities and explanations.

25. Going, Going, Gone Once

Activity

Objectives
- to examine evaporation
- to learn about contour mapping

Materials
- chalk, paper towels

Observe puddles on your playground. Discuss what happens to the water. **What helped make the water disappear?** Discuss how the size of the puddle changes.

Following a rain (and preferably after the sun comes out), have children use chalk to trace the outline of a puddle on the playground (concrete or asphalt areas). A few hours later, or even the next day, have children trace the outline again. Do this several times, until the puddle disappears (i.e., evaporates). The pattern that remains is a contour map, showing slight changes in the height of the ground. Water runs from higher ground and collects in lower-lying areas.

To further experiment with evaporation, use different variables. Have children wet paper towels and place these in different places. Some should be in a sunny areas, with others in the shade; also, spread some out flat and roll some into balls. Some children might hang their paper towels "out to dry," just like some people hang clothes. **Which paper towels dried fastest? What are the "best" conditions for increased evaporation or drying? Can children use this information to help dry out wet clothes, wet gloves, and wet towels, that might be at home or at school?**

26. Going, Going, Gone Twice

Experiment

Objectives
• to discover properties of ice
• to experiment with melting

Materials
• ice cubes, plastic bag, ice chest

Hold an ice melting contest. Fill an ice chest with small plastic bags, each containing a single ice cube (of similar size and shape). Then distribute to children and have them try to be the first to melt their ice cube. **What are some of the ways the children did this?** Some may have put it under their arm, blown on it, placed it by a sunny window, broke it into pieces, or placed it into a bowl of warm water. Discuss results and decide upon the variables that affect how fast an ice cube will melt. Basically, the more heat that is added to the ice, and the greater the area exposed to the heat, the faster the melting occurs.

27. Ice is Nice

Experiment

Objectives
- to discover more properties of ice
- to experiment with freezing

Materials
- ice cubes, index cards or poster board, plastic bag, freezer, and margarine tubs, plastic milk or soda bottles, or other clear cylindrical plastic containers

To demonstrate what happens to water when it freezes, fill a plastic container nearly full with water. Use a permanent marker or tape to mark the water level. Either place the container in a freezer, or out of doors in a shady location, if it's cold enough. Have children predict what will happen. Most will state that the water will freeze. After it freezes, examine the container. **Did the ice take up more or less space than the water?** Have children examine the ice and see if they can find out why the ice took up more space (air bubbles). You may want to repeat the experiment using a more rigid plastic container (e.g., food storage type, or a two to four inch high plastic specimen container).

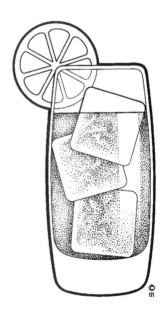

28. Icebergs

Activity

Objectives
• to discover more properties of ice
• to examine icebergs

Materials
• ice cubes, margarine tubs, index cards or poster board, plastic bag, freezer, plastic see-through sweater box container

Tell the children that you are going to place a large piece of ice (from Activity 27) into a container filled with cold water. Have children predict what will happen to the ice. **Will it sink or float?** Some may think it will sink to the bottom because it is bigger than an ice cube. Actually, any size piece of ice will float in water because the ice contains air bubbles and is less dense than the water it displaces. However, about 90% of the ice cube will float below the surface of the water. This is what happens with icebergs.

Have children view these icebergs from the side. **Can they estimate how much of the iceberg is below water?** Typical answers for very young children might be "lots;" older children can use fractions or descriptors. Have older children devise a method for measuring the amount of ice above and below the surface of the water.

29. Ice to Vapor

Experiment

Objectives
- to discover more properties of ice
- to experiment with sublimation

Materials
- ice cubes, index cards or poster board

Ice and snow can also change from a solid to a gas. It does this without ever melting. This is called sublimation (see Activity 24). To observe this, have children put an ice cube or snowball on an index card or small piece of poster board and place in a location where the temperature is below freezing. They can trace the shape of the ice cube or snowball on the card or measure and record its dimensions. In colder climates, they can use a shady location out of doors. In any climate or season, they can use a frost-free freezer. Have them observe the ice cube on several consecutive days (or even several weeks) and have them retrace its shape and/or remeasure it. **What happens to the size of the cube? Do children observe any melting? For example, did the card get wet?** In a frost-free freezer, the snowball and the ice cube will sublimate (i.e. change from a solid into a gas -- water vapor -- without first melting).

Also have children place a snowball or ice cube into a zip-lock sandwich bag onto a paper plate or piece of paper before putting it in the freezer. Compare and contrast what happens overnight, over a few nights, and over several weeks. Have them retrace shapes and/or remeasure.

In class, have children share the methods they used for measuring (both standard and non-standard).

Suggested reading: *The Snowy Day*

LEADER CHALLENGE: Have children look skyward to find airplane trails (these are called CONTRAILS, short for CONdensation TRAILS). On some days, these contrails, made of ice crystals, last for a long time; on other days, they sublimate almost as soon as they form.

Ask the children what cloud type they think contrails are.

Wind

30. Air, Air, Everywhere

Demonstration

Objective • to recognize the presence of air

Materials • plastic see-through sweater box container, small clear container, water, paper towels

Before beginning this demonstration, fill a plastic sweater box about three-fourths full of water.

Hold up a clear, small, empty container and ask children what is in it. Be prepared for answers like "nothing." Stuff a dry paper towel completely inside the container and tell the children that you can put this container under water (inside the sweater box) and that the towel will stay dry. Some children may tell you that the towel will get wet. Poll children for their opinions and post results.

Submerge the container, holding it upside down. Remove the container and take out the paper towel. The towel will be dry, other than a few spots of water from contact with the lip of the container. Have several children verify for the class that the towel is dry. Discuss with children that air was inside the container. To show them the air, again submerge the container, this time without a paper towel inside. Slowly tilt the glass. Children will see air bubbles escape.

Discuss with children places that air exists (e.g., in ice cubes, between snowflakes, in a balloon, in bicycle tires, in our atmosphere). Tell them that we breathe air and that air is also in our lungs and inside our bodies. Tell them that air contains oxygen and other gases. Sometimes the air contains pollutants, too. Some children may have shown their understanding of air in Activity 14.

31. Up, Up and Away

Demonstration

Objective • to realize that air has pressure and exerts a force

Materials • 11" or larger diameter balloons, small books

Blow up and hold the neck of a balloon. Ask children what is in the balloon. Quickly release the neck of the balloon, letting a small amount of air out. Have children describe the sound they hear (as the rush of air escapes). Ask children why the air tries to escape from the balloon. (It's because the air pressure is greater inside the balloon - you forced air inside when you blew up the balloon.)

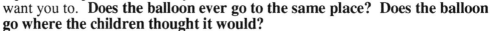

Blow up a balloon and tell the children that you are going to release it. **What is their reaction?** Release the balloon and let it fly around. Have children describe its motion. Discuss the force of air coming out of the balloon and how it pushed the balloon in the opposite direction.

Tell children that you are going to repeat the demonstration, releasing the balloon exactly as you did before (i.e. in the same direction, with about the same amount of air inside the balloon). Have them predict where the balloon will go and ask them to stand in that place. Release the balloon. You can repeat the experiment as many times as the children want you to. **Does the balloon ever go to the same place? Does the balloon go where the children thought it would?**

Tell children that you are going to use a balloon to lift a book. Children may tell you that it can't happen. At the edge of a table, place a balloon underneath a book, leaving the neck sticking out. Blow up the balloon. The force of the air inside the balloon is greater than the weight of the book. Try this with several books piled atop each other, too. **How many books can you lift this way?**

You can also lift a chair. Place a small chair upside down on a table with a balloon sticking out, as you did for the book(s). Repeat the experiment. **What other things are lifted by air in this manner?** The list includes car and bike tires, air mattresses, hot air balloons, and swimming tubes.

32. Batter Up!

Activity

Objective • to realize the many types of forces associated with wind

Materials • balloons, parachute, paper towel rolls, markers, crayons, chairs

Blow up enough balloons so that each child has one. Lay out a parachute (or a large sheet) and place all balloons near its center. Have children sit in a circle holding the handles of the parachute. If there are not enough handles, some children can hold the edge of the parachute. Have the children gently lift and lower the parachute. Watch what happens to the balloons! Discuss the forces that made the balloons move.

Color or decorate paper towel rolls. Have each child select a balloon. Using the decorated balloon bats, children should try to keep a balloon in the air and not allow it to touch the ground. Have them try this individually and then with a partner. **How long can they keep a balloon in the air?**

Have a balloon golf relay race. Divide the class into two teams, one on each side of the room. Have one child use their decorated balloon bat to putt their balloon around two chairs and to the other team. The first child on the other team can start putting the balloon back to the first team, again going around the two chairs. Continue until all children have had a chance to putt. You can make the "obstacle course" as difficult as you and the children want it to be.

Was it easy to move the balloons with the bats? Did the balloons go where you wanted them to? What force(s) might have acted on your balloons? (air currents, friction, the bat did not always hit the balloon directly - sometimes it forced the balloon to go in another direction, or maybe the balloon bumped into something and bounced off).

Discuss sports or activities that involve balls or other objects that move through the air (e.g., golf, baseball, football, Frisbees™). These same forces affect how these balls move through the air.

33. Balloon Rockets

Demonstration

Objective • to continue to study forces associated with wind

Materials • balloons, string, straws, masking tape

Build a balloon rocket track as shown below. Place a straw on a length of string that spans your classroom. Attach the ends to opposite walls at an accessible height for the children. Be sure that there are no obstacles in the way.

Blow up a balloon, holding the neck to keep the air trapped. With the straw at one end of the string, attach the balloon to the straw using masking tape. Countdown to blastoff (5...4...3...) and release the balloon. How far did the balloon go? Repeat several times, changing variables. For example, you might add more or less air to the balloon, or you might use balloons of different sizes or shapes. You can also change the height of one end of the string so the balloon goes up or down at an angle. Discuss results. Discuss how this could affect the space shuttle, airplanes, and other things that fly.

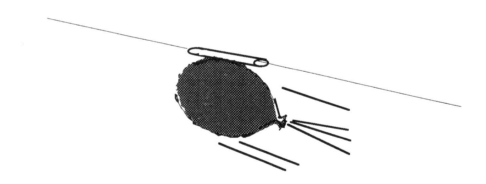

34. Balloon Rocket Race

Activity

Objective • to continue to study forces

Materials • balloons, string, straws, masking tape

Divide the class into teams of four to six and make balloon tracks (as described in Activity 33) for each team. Have half of each team positioned at each end of their string. Give each child their own balloon and a piece of masking tape. Set up a relay race.

All children should blow up and hold the neck of their balloon. For younger ages, you may want to blow up balloons and give them to the children to hold. Children should attach a loop of masking tape on the side of their balloon. At a given signal, the race begins. The child at position one attaches his/her balloon to the straw using the masking tape loop and releases the neck of the balloon. If the balloon does not reach the other end of the track, the child must blow up the balloon again, reattach it to the straw and release it. Once the balloon reaches the other end of the track, the child at position two, removes the first balloon and attaches his/hers and sends the balloon back in the opposite direction. The race continues until all children have launched their balloon rockets and completed their leg of the race. The team completing this first is the winner.

Discuss strategies, if any, that were used in launching the balloons. **Where was the masking tape placed? Was the balloon aimed a certain way before it was released? Did everyone blow up their balloon as much?**

35. Flighty Activities

Activity

Objective • to discover more about air and its forces

Materials • sheets of 8 1/2" x 11" paper, paper clips, data sheets

Hold the short edge of a piece of paper and extend the paper horizontally away from you. Then relax your grip enough so the other edge of the paper drops down. Tell your class that you can make the other end of the paper lift up without moving your hands. Be prepared for responses including "it can't be done."

Blow across the top of the paper, gently at first. The paper should start to rise. Hint: practice this before trying it with your class. Have children comment about what they think happened. (This is Bernoulli's principle. As air moves past the top of the paper, the air pressure lowers. The higher pressure under the paper lifts it.) Tell children that this lift is one of the factors that allows airplanes to fly. Have children (especially in older grades) try it themselves.

Make and fly paper airplanes. Using the sketches below as a guide, fold a piece of paper in half lengthwise. Fold down corners at one end. Then fold the entire length in half on each side. Put each child's name on the wing of his/her plane. Children can also decorate their planes. Attach a paper clip at the front of the plane. Then fly the plane. Refer also to *Flights of Imagination* and *More Mudpies To Magnets*.

Have children line up in groups of 4 to 6 at one end of a large room (e.g., gymnasium or multi-purpose room) and fly their planes toward the opposite end. Mark or measure the room so standard distances (e.g., 3 feet, 10 feet) are shown using masking tape. For all flights, measure (using a ruler, yard or meter stick, or tape measure) and post flight length. On a separate poster keep track of the longest three flights. Be sure to always note the owner of the planes. Allow children to fly their planes several times, adding this data to

that originally posted. After all flights have been completed, have children write flight data for their group on individual data sheets. As in all "competitions," encourage children to compete against themselves (personal best) as well as competing with others.

Which planes flew the furthest? Did the same plane always fly the furthest? Assuming that each plane was built the same way, what are some of the other variables which could account for differences in the length of the flights?

Have children brainstorm how to redesign their airplanes to make them fly further. **For example, could the paper clip weight be moved to the center or back of the plane? Could more weight be added? Should heavier or lighter paper be used? Should larger or smaller sheets of paper be used? Should the wings be bent differently to create additional lift? Should the planes be thrown in a different way? Would the planes fly further if they were thrown from the opposite end of the room? Would the planes fly further if they were flown on the playground?**

Following the brainstorming activity, have children redesign their planes. Repeat the activity as described above tabulating data for flights #4 - 6. Compare and contrast results. Again ask questions like **Which planes flew the furthest?** Complete the SCIENTIFIC METHOD DATA SHEET on Page 61.

_____ _____
         name   date

"FLIGHT DISTANCES"
INITIAL DESIGN - DATA SHEET

name	flight #1	flight #2	flight #3

"FLIGHT DISTANCES"
MODIFIED DESIGN - DATA SHEET

name	flight #4	flight #5	flight #6

© 1996...H. Michael Mogil and Barbara G. Levine, 1522 Baylor Avenue, Rockville, MD 20850

_____ _____
            name   date

SCIENTIFIC METHOD DATA SHEET

Method Used: _____

_____

_____

_____

_____

_____

_____

Observations: _____

_____

_____

_____

_____

_____

_____

Conclusions: _____

_____

_____

_____

_____

_____

_____

© 1996...H. Michael Mogil and Barbara G. Levine, 1522 Baylor Avenue, Rockville, MD 20850

36. Where the Wind Comes from (Building a wind vane)

Activity

Objectives
- to observe the wind
- to estimate the direction from which it blows

Materials
- poster board (fluorescent colors preferred), unsharpened colorful pencils with eraser, scissors, push pins (already placed into eraser for safety, straws cut in half, stapler, template

Go outside and have children observe the wind. Ask them how they can tell that the air is moving. (Their answers should reflect how the wind affects them and their environment - trees sway, leaves move around, flags wave, I feel cold).

Brainstorm about the ways that people artificially create wind. Answers could include all types of fans (box, ceiling, hand-held), air conditioners and hair dryers. Children can also blow air, creating their own wind. See Activity 8 in which children blow clouds of shaving cream.

Meteorologists describe winds by the direction FROM WHICH they come. Although this usually means a compass direction, it does not have to be. **Have you ever heard of a sea breeze (a breeze that blows from the sea)?** Your children can estimate the direction from which the wind is blowing either way. Older children can use compass directions, while younger children can use landmarks (e.g., the wind is blowing from the tall oak tree).

The WIND VANE is used to measure wind direction. Children may have seen these atop roofs. Often they have an animal along with the letters of the four main compass directions. Make wind vanes using the following procedure:

- Place a push pin in center of the straw and push partly into pencil eraser. Have children try to spin the straw. It probably won't work well because the straw rubs against the eraser and the pin head.

- Have children flatten straw near the pin by stapling on each side of the pin hole. Don't staple too close to the push pin hole. Replace the straw and test to see if the vane spins freely. Adjust as needed.
- Using patterns shown in the template below, cut out an arrow head and arrow tail from poster board. Try different sized heads and tails.
- With straw attached to the pencil, cut a small vertical slit at each end of a straw. Insert arrowhead and tail into each slit and staple twice.

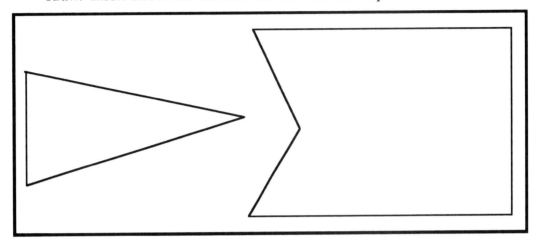

Before using your wind vane, test it with an electric fan or hair dryer located several feet from the wind vane. **Ask children which way they think the wind vane will point (into the wind or with the wind?).** The arrowhead should point toward the source of the wind, as long as the tail is bigger than the arrowhead. To prove this, make a wind vane with a head larger than a tail. Discuss why this happens with the children.

Take the wind vanes outside and hold them in an upright position to measure the wind direction on several days. Compare your observation of wind direction with that reported on local television or radio weather reports. Remember that the wind direction is the direction **from which** the wind blows.

Does each child's wind vane point in the same direction? Does the wind always blow from the same direction? Is a north wind always cold? When the wind blows from the east does it always rain or is it more likely to rain? Add this weather variable to the class' weather calendar (Activity 3).

> LEADER CHALLENGE: Have children go home and find objects that can be blown around by the wind. These could include feathers, pieces of paper, Styrofoam cups, leaves, branches, and seed pods. Children should bring these into class.
>
> Then working in groups, have children order the objects by their ability to be blown around by the wind. What are some of the characteristics of the objects which are most easily moved?

Suggested reading: *Gilberto and the Wind*

37. Blowing in the Wind

Experiment

Objective • to observe wind forces

Materials • two identical cereal boxes (one empty and one full).

Place the two cereal boxes at the edge of a table, with the bottoms of the boxes facing the children. Ask for two volunteers. Have each child attempt to blow over their cereal box. Repeat the experiment by having the children change places. Note that the children should not touch their boxes and you should not tell them anything about the contents (or lack of contents) of the boxes. Discuss what happens. (Nothing should happen because the wind from the children does not exert enough force on the small area of the bottom of the cereal box).

Turn the boxes so that the fronts face the same two children. Repeat the experiment. One of the boxes should be blown over. Upright that box and have the children change places. Repeat the experiment. Discuss why only one box blew over. (The force of the wind created by the child(ren) was sufficient to blow over the empty cereal box because of the larger surface area. This force was not enough to overcome the weight of the full cereal box). Since more wind will be needed to blow over the full box, have children predict how many more blowers will be needed to blow over the full box. Then try it, first with two children, increasing the numbers until the box falls over. Note: it may not be possible to blow over the box with very young children and it may not be possible to get enough older children close enough to the front of the box to blow it over. However, it should be fun trying!

38. Air and Water Streams

> **Demonstration**

Objective • to observe the behavior of wind

Materials • coffee can, clay/play dough, birthday candle, matches, cookie sheet, sweater box

Activity -- Put a birthday candle into clay or play dough. Place a coffee can between the candle and the edge of the table. Ask for volunteers - **"Who thinks they can blow out the candle that's behind the coffee can?"** Select a volunteer and position him/her in front of the coffee can so that they cannot see the candle. Light the candle. Have the other children gather where they can watch the candle. Ask the volunteer to blow. Observe what happens.

Was the volunteer able to blow out the candle? Did the "wind" go through the can? over the can? or around the can?

Have children go outside and look for places where the wind blows strongly around obstacles (e.g., buildings). This is called a "wind tunnel" effect. Have them find places where the wind isn't as strong (e.g., behind rows of bushes, or trees). These places are called "shelters."

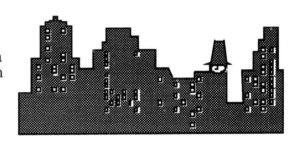

After the children come back inside, ask them what the building they are in is called. Remind them that it protects them from the wind and is also called a "shelter." Discuss where different animals build their homes. Birds, for example, make nests in the leafy parts of trees. One reason for this is shelter from the wind. Other animals build their homes inside trees, in caves, and in the ground.

Demonstration #1 -- To demonstrate that air and water have a similar properties concerning their movement, create a miniature stream table using a cookie sheet and a sweater box. Tip the cookie sheet slightly and place a fist-sized rock or piece of clay about 1/3 from the top. Gently pour water so that races past the rock and collects in the sweater box. Observe what happens. Compare this to air flow around the coffee can.

Demonstration #2 -- Repeat the experiment using two rocks or pieces of clay next to each other. **How does the water flow between the rocks?**

Suggested reading: *A Letter to Amy*

39. Wind Speed Indicator (Building an anemometer)

Demonstration

Objective • to observe, estimate and measure wind speed

Materials • 2 table tennis balls (cut in half), flower vial, 3 bamboo sticks, 2 1/2" Styrofoam ball, X-ACTO™ knife or single edged razor blade, permanent marker

Make an anemometer following the procedure described below and use the anemometer to estimate or measure wind speed.

- Carefully cut table tennis balls in half using an X-ACTO™ knife or razor blade.
- Color one table tennis ball half a different color using a permanent marking pen (you can also purchase table tennis balls that have different colors or designs).
- Assemble table tennis ball halves (called cups), Styrofoam ball and bamboo sticks as shown in diagram below. This is your anemometer or wind speed indicator.
- Place anemometer into a flower vial.
- Align table tennis ball cups so they all face in the same direction as they rotate.
- Have a child hold the anemometer as you aim a hair dryer toward it blowing at low speed. Repeat blowing at high speed. Have children look for differences in how fast the anemometer spins.

Take the anemometer outside to an open area on a windy day. Have a child count the number of revolutions the different colored cup makes in 30 seconds. The number of revolutions provides a relative measure of wind speed; the more revolutions, the windier it is. Have children compare how they felt the wind (or how trees moved or flags waved) to the number of revolutions made by the anemometer. This will help them create their own wind scale.

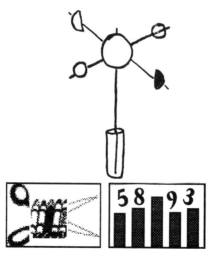

40. Pinwheel Mania

Activity

Objective • to create a wind catching device

Materials • Poster board, straws, paper fasteners, stapler, markers or crayons, hole punch, hole reinforcers

Make a pinwheel following the procedure described below and let children "play with the wind."

- Provide each child a 6" square piece of poster board
- Have children color and decorate both sides. We recommend using bright colors and wild patterns!!
- Pencil an X from corner to corner and then make three inch cuts toward the center from each corner.
- Punch a hole in the center and place hole reinforcers on both sides.
- Begin at any corner and loop the pointed corner piece nearly to the center of the paper. Staple the pointed end down. Repeat for every corner.
- Using a hole punch make a hole near one end of the straw.
- Attach the pinwheel to the straw with a paper fastener.
- Take pinwheels outside on a windy day and let children experiment with them.

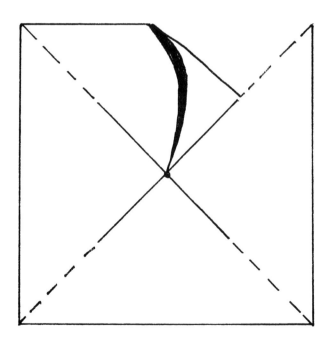

41. Make Your Own Wind Machine

> **Activity**

Objective • to invent a wind catching machine

Materials • Straws, Styrofoam balls, coffee stirrers, tooth picks, small and large paper/plastic cups, plastic caps, egg cartons, Styrofoam packing material and trays, pipe cleaners, small pieces of fabric, paper, construction paper, tissue paper, small boxes (e.g., individual cereal), clay, scissors, cold glue gun

Provide a collection of materials (such as those described above) to the children and have them invent a machine that will move by the wind. There are no "rules" for this activity. Some children may need help getting started; others may require adult assistance building their device. The bottom line is to let imaginations fly freely!

42. Tornado!!!

Demonstration

Objective • to observe wind forces

Materials • 2-liter plastic soda bottle, water, liquid dishwashing detergent, miscellaneous small objects, tennis ball container, stirrer, golf ball

Demonstration #1 -- fill the plastic soda bottle two-thirds full of water. Add a single drop of dishwashing detergent. Cap bottle tightly and shake. If you have too many bubbles, dilute your solution by filling the bottle past overflow so that some water and lots of bubbles are expelled. Pour off some water. Test for proper amount of bubbles and repeat diluting procedure until you have only a small amount of bubbles.

Invert and hold the bottle vertically at both the top and bottom; then move your hands in a small circle around a vertical axis. It doesn't matter which way you move your hands as long as they move together. As you move the bottle, you should see a funnel-shaped pattern appear. The faster you move your hands (which simulates increasingly faster wind flowing in a spinning manner), the more pronounced the pattern becomes. TORNADO!

Stop moving you hands once the tornado appears. Hold the bottle where children can see the tornado. Once the tornado weakens (the spinning slows down), move your hands in a small circle again to recharge the tornado.

Add small objects (e.g., Lego™ pieces, small folded aluminum foil sheets) to show how things spin around a tornado and how they are pulled into it.

Demonstration #2 - Fill a tennis ball container about 2/3 full of water. Place a golf ball in the container. Stir the water rapidly, but only at the top inch or two. Watch the golf ball be "pulled up" by the whirling motion of the tornado.

Activity -- Have children draw or paint pictures showing tornadoes in action. Alternatively, have them draw a "friendly" tornado.

  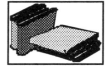 

Has anyone ever seen a tornado? Where? What did the tornado look like? What did it sound like? Have they ever seen the tornado in the movie "The Wizard of Oz"? Why do you think tornadoes are dangerous? What are some tornado safety rules?

> LEADER CHALLENGE: Have each child bring in a small bottle which can be resealed (e.g., spice, soda) and make individual "pet" tornadoes.

Suggested reading: *Tornado Alert*

Sun

43. Sun Prints

Activity

Objective • to observe one effect of sunlight

Materials • colored construction paper, other pieces of paper, scissors, miscellaneous objects

Have children trace their hand print onto a piece of paper and cut it out. Place the cut out hand shape onto a piece of construction paper (dark blue works best, but children should try various colors) and put both outside in a sunny location. Weigh down the cut out hand shape using rocks or other heavy objects. Leave outside for at least an hour.

Bring inside and remove the cut out hand shape. Have children examine the blue paper and discuss what the they observe.

Alternative activities include making collage prints using various objects found around the classroom, school or neighborhood (e.g., leaves, rocks, scissors, cut out shapes, stencils, flowers). Children can also make cloud prints using stencil cloud shapes. This will produce unfaded dark blue clouds against a faded blue background.

> LEADER CHALLENGE: Encourage children to talk about the effects of sunlight fading with their family. Clothing, fabrics on furniture and drapes, and paint are among the objects that fade in sunlight.

44. Sunbathing

Activity

Objective • to observe one effect of sunlight

Materials • colored construction paper, scissors, miscellaneous objects

Repeat Activity 43 by changing several variables. Place several **sheets of the same color construction paper outside on a cloudy day and a sunny day.** Each hour bring one sheet inside. Mark the time on the sheet and place the sheet upside down on a table. At the end of the day, turn all sheets over and compare the amount of color change to the amount of exposure to the sun. Discuss how this relates to the exposure of human skin to sunlight. Ask children if anyone has ever had a bad sunburn. Talk about how the sunburn felt. Discuss what happened to their skin. Be sure children understand sun safety.

Repeat the activity using sets of different colored construction paper. **Were the results similar or different?** Discuss. This is an opportunity to talk about how the sun might affect people with different colored skin.

> LEADER CHALLENGE: Exposure of human skin to sunlight is a very important health and safety issue. Encourage children to share information about sun safety with other family members. Although sun safety is paramount in the summer when children are out of doors and/or at the beach, it can be a significant factor in the winter, too, especially in mountain areas and when the ground is snow-covered.

Suggested reading: *People*

45. Shadows

Activity

Objective • to observe how shadows change

Materials • colored chalk

On a sunny day, have children scatter around a blacktop or concrete area with their backs to the sun. Have them partner and take turns tracing each others shadows. Also trace the footprints of each child and include these in the shadow. Have the children write their names next to their shadow.

Do other activities (e.g., Activities 46 and/or 47) and return in 30 minutes or so. Have children stand in their original footprints and observe their new shadows. **Can they fit back into their original shadow by either relocating their feet or changing the way they stood?** Brainstorm with children what might have happened. (Typical responses might be that the sun moved or the shadows walked. Actually, it was the Earth spinning on its axis -- the cause of day and night -- that changed the children's position relative to the sun.) See also Activities 51 and 52.

Repeat the activity several times during the day. **What else do the children notice?** They might see that their shadows get small and then big - a human sundial; they might also notice that the tips of their head lie along a straight line - that line is oriented west to east, making this a great way to reinforce compass directions.

If it doesn't rain, and the shadows are not disturbed, have children revisit their shadows at the same time on succeeding days.

Suggested reading: *What Makes Day and Night*

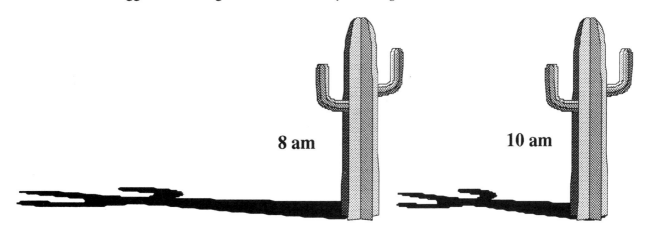

46. Shadow Tag

Activity

Objective • to have fun with shadows

Materials • none

On a sunny day, preferably on a grassy area, play shadow tag. Choose one child to be "it." That child must tag the shadow of a classmate by stepping on it. Then that child becomes "it." A child is safe if he/she puts themselves totally in a shadow, because then the child does not have a shadow!

47. Take a Shadow Hike

Activity

Objective • to observe shadows of common objects

Materials • paper, pencil

On a sunny day, take children on a hike to hunt for shadows. Observe the different shadows, noting their shapes. Have children record and/or sketch the sizes of various shadows. Have them use non-standard measurements to document their shadows (e.g., 25 hands tall or 5 footsteps wide). **Can children fit into the shadow of a particular tree or building?**

To observe how shadows change, take one hike early in the morning and then another in the early afternoon. Try to observe and measure the shadows of the same objects.

48. Cloud Shadows

| Activity |

Objective • to observe cloud shadows

Materials • none

On a day with cumulus clouds, have children watch how cloud shadows move across the area. If tall buildings, hills, mountains, or large open areas are nearby, they'll be able to see the shadows more easily.

As the cloud passes between you and the sun, it blocks the sunlight, thereby creating a shadow. Thicker clouds usually produce darker shadows. The bottom of thunderstorm clouds can be very dark! This is because the cloud's shadow is also being cast on the bottom of the cloud. On a day that is completely cloudy (i.e., no sunshine), you shouldn't find any shadows.

Use sky and cloud windows (Activity 9) to measure the darkness of the bottom of clouds.

On a sunny day, bring sheets of copying paper outside and divide the class into pairs. Each pair should do the following. Have one child hold a single piece of paper up to the sunlight. **Looking down, does it cast a shadow on the ground? Looking up, can the other child see the light coming through the paper?** Repeat, adding additional sheets of paper. **How many pieces of paper does it take before the light cannot be seen?** Compare the thickness of the papers to the thickness of clouds. If the weather outside does not cooperate, the demonstration can be done using a powerful flashlight, a lamp, or an overhead projector. *Be sure children are careful with their eyes around these light sources.*

49. Groundhog Day

Discussion

Objective • to learn about weather folklore

Materials • none

Groundhog Day comes in early February and offers a great opportunity to introduce weather folklore to the children. Tell about the groundhog legend and how the groundhog is supposed to predict the weather for the end of winter.

The most famous groundhog lives in Punxsutawney, Pennsylvania. **How far is Punxsutawney from where you are?** When the weather is cloudy in Punxsutawney, does that mean it is cloudy everywhere? Have children become groundhogs and observe their shadows. Compare their forecast for the next six weeks of winter with that of Punxsutawney Phil and any other groundhogs that are in the news. Are there any other animals in your area that allegedly can predict the weather? Recent news stories have talked about woolly caterpillars in North Carolina and a snake in Arizona that are weather forecasters. They should also keep a groundhog weather calendar to verify their forecasts.

Tell about other weather proverbs or sayings that involve animals or insects. **Do the children think these are true?** The book *Weather Proverbs* by George D. Freier is a good teacher's reference. Here are a few proverbs for short-term forecasting to get you started:

- If the rooster crows on going to bed, you may rise with a watery head.
- Everything is lovely when the goose honks high.
- When birds huddle at the top of a chimney top, it is a sign of cold weather.
- If a dog pulls his feet up high while walking, a change in the weather is coming.
- Frogs croaking in the lagoon, means that rain will come real soon.
- Spiders enlarge and repair their webs before bad weather.
- Bees never get caught in a rain.
- Expect stormy weather when ants travel in a straight line; when they scatter all over, the weather is fine.

Suggested reading: *Geoffrey Groundhog Predicts The Weather*

 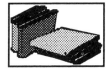

50. Plants

Experiment

Objective • to study the effect of sunlight on plants

Materials • 4 to 6 small bedding plants that require direct or full sun

Place half the plants in a sunny location and the other half in a dark location. Water plants lightly when the soil becomes dry to the touch, but be sure to control the amount of water used (a key variable). Have children observe plants and create a picture journal. Compare plant height, number of leaves, leaf color, and other plant characteristics among plants in sunny and dark locations.

You may want to repeat the experiment using plants that require shade. Compare and contrast the outcomes of the two experiments.

Change other variables (e.g., sunlight versus fluorescent light, window on south versus north side of building, type of soil used). Compare and contrast results.

51. Sundials

Activity

Objective • to use changes in shadows to tell time

Materials • markers, crayons, clay / play dough, paper plates, straws

Have each child make a sundial by placing a ball of modeling clay in the center of a paper plate. Press down gently on the ball so that the bottom becomes flat. Insert a straw into the ball of clay so that the straw is pointing straight up. Trace the location of the clay ball on the plate. Put child's name on each plate.

On a sunny day (that isn't too windy), place plates outside in a sunny location. Be certain to place the clay ball in its proper place. Trace the shadow of the straw at hourly intervals throughout the day, labeling each shadow with the time. At the end of the day each child will have his/her own sundial. Be sure to mark a direction from a point on the edge of each plate to a playground object. This will help align the sundial on other days.

Place the sundials on the playground on other sunny days, aligning them as described above. Have children use the sundials to tell time. Compare the time shown on the sundial with that shown on a watch.

If it is difficult to do this activity out of doors, it could be done indoors in an area which receives sunlight for many hours (e.g., a southward facing lobby).

Alternatively, repeat Activity 45 hourly to create a human sundial. Since the children will be retracing their shadows several times, be sure to leave enough space between children so that the shadows from different children don't overlap.

52. Day and Night

> **Activity**

Objective • to experience day and night

Materials • none

Have children pretend that they are the Earth, spinning around the sun (played by an adult leader). Form a circle around the sun and have each child spin around (their vertical axis), thereby rotating, until the adult leader shouts "freeze." Children should stop spinning and see where they are relative to the leader. Those facing the adult leader are "days" and those facing away from the leader are in the Earth's shadow and are "night." If the child is sideways, he/she is between day and night; this means either sunrise or sunset.

This is an important concept. We typically say that the sun rises and sets. In reality, it is the Earth that is spinning, creating what appears to be a rising and setting sun.

Suggested reading: *What Makes Day and Night*

 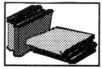 

53. A Solar Web

Activity

Objective • to show how the sun is related to everything on Earth

Materials • index cards, markers, crayons, string

Have each child name one object that they see in their playground environment. Have each child draw and label their object on its own index card. Each child holds onto his/her card.

Have the children sit in a circle. An adult leader should place an index card with a SUN in the center of the room. Have one child place their index card near the SUN card. If their object relates to the sun, they should place a piece of string between their card and the SUN card. Cut pieces of string as each card is placed in the web.

A second child should see if their object relates to either card. If it does, they should place a piece of string between their card and one of the cards on the floor. Each succeeding child should follow this lead, linking their card to only one other card. Discuss the relationships that the final web shows.

Have each child return to the web and try to match their card with another card. After all children have had a chance, discuss the relationships of the new, more complicated, web. Show children other relationships (webs) which they may have not have discovered.

What happens to the web if the sun is removed? Have each child remove their string which relates to the sun. You can reconstruct the web and remove a different object to see what happens when something else is removed from an environmental web. Have children discuss what might happen to the environment if other objects were removed (e.g., no birds, no soil, no water, no people).

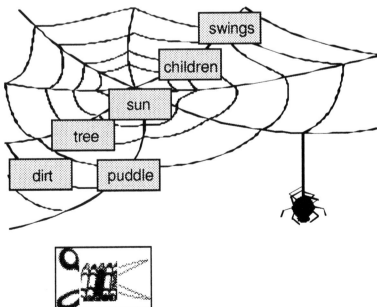

54. Rainbows

> **Demonstration**

Objective • to show how all of the colors are contained in sunlight

Materials • misters, Rainbow Glasses™, poster board, 2 sheets of paper, tape, prism, cut crystal

Do this activity on a sunny day. Fill several misters with water and go outside with your class. Have children face away from the sun. Have adult leaders stand to the side of small groups of children and spray the mister across the children's path. Sunlight shining through this mist should create a rainbow-like display. Depending upon the time of day (and hence the angle that the sunlight is striking the drops), the children may have to move to different positions to see the rainbow.

The rainbow is created when sunlight passes through raindrops and is reflected off the inside surface of the drops. As the sunlight passes from air into the drop and back out, it is bent slightly. The bending is different for each of the colors making up sunlight. Thus, the journey through the raindrop acts to separate the colors. The order of the colors is Red - Orange - Yellow - Green - Blue - Violet. Red is on the outside of the rainbow.

You can also use a prism with an intense light source (e.g., slide or overhead projector), a cut crystal on a string in bright sunlight, or a diffraction grating. The latter is contained in Rainbow Glasses™. Cut out a small window in a small piece of poster board and cut out one of the lenses (with frame) from Rainbow Glasses™. Tape the lens into the poster board window.

Using an overhead projector, in a darkened room, lay two pieces of 8 1/2" x 11 paper on the projection glass. Leave a very narrow vertical space between the two papers. Then hold the poster board window by the projection lens. You should create a vivid rainbow. Change the width of the vertical space between the sheets of paper and you will change the size of the color bars in the rainbow.

When have you seen rainbows? In which season? At what time of day?
Most often rainbows are seen in spring and summer in most parts of the United States. Most are seen late in the afternoon, to the east, usually after a shower or thunderstorm. Hawaii is called the Rainbow State because it has so many rainbows; these occur because it may rain over the mountains, but be sunny at the beaches.

Activity 66 provides another opportunity to create rainbows.

55. The Story of the Rainbow

Activity

Objective • to understand rainbows

Materials • Rainbow Glasses™, assorted colored construction paper, crayons, markers, paint

Activity -- Read "The Story of the Rainbow" on the next page. Adapt the text, as appropriate. Have pairs of children act out the parts of the colors and the rain. Children should either wear or hold the color that they are portraying. Either use colored construction paper or have children paint or color paper to represent colors of the rainbow. When the colors join hands at the end of the story, have them line up according to ROY G BV to represent the actual arrangement of colors seen in a rainbow. RED is on the outside of the primary rainbow. So the children would line up as ROY G BV - VB G YOR to show both ends of the rainbow. ROY G BV is a mnemonic for
RED --- ORANGE --- YELLOW --- GREEN --- BLUE --- VIOLET.

Give each child a pair of Rainbow Glasses™. Adjust side pieces as needed. Have children look around the room. They should see interesting color patterns. Look at a light in the room. A fluorescent light will give color bands. A round bulb will give a radial pattern. *As always, remind children NOT to look directly at the sun!*

The Story of the Rainbow

Once upon a time, all the colors in the world started to quarrel; each claimed to be the best, the most important, the most useful, and everyone's favorite.

Green said: "Clearly, I am the most important. I am the sign of life and of hope. I was chosen for grass, trees and leaves. Without me all the animals would die. Look out over the countryside and you will see that I am in the majority."

Blue interrupted: "You only think about the earth, but consider the sky and the sea. It is water that is the basis of life and is drawn up by the clouds from the blue sea. The sky gives space, peace and serenity. Without me you would all be nothing but busybodies."

Yellow chuckled: "You are all so serious. I bring laughter and warmth into the world. The sun is yellow. Every time you look at yellow roses, sunflowers, marigolds and even the dandelion, the whole world starts to smile. Without me there would be no fun."

Orange started to blow its trumpet: "I am the color of health and strength. I may be scarce, but I am precious for I serve the inner needs of human life. I carry all the most important vitamins. Think of carrots, pumpkins, and oranges. I don't hang around much, but when I fill the sky at sunrise or sunset, my beauty is so striking that no one gives a thought to any of you."

Red could stand it no longer and shouted out: "I am the ruler of you all! I bring blood, life's blood! I am the color of danger and of bravery. I am willing to fight for a cause. Without me the earth be as empty as the moon. I am the color of love; the red rose, the poinsettia and the poppy."

Violet rose up to full height, standing with great pomp and authority: "I am the color of royalty and power. Kings, chiefs, and leaders have always chosen me for I am a sign of authority and wisdom. People do not question me; they listen and obey."

And so the colors went on boasting, each convinced that they were the best. Their quarreling became louder and louder. Suddenly, there was a startling flash of brilliant white lightning; thunder rolled and boomed. Rain started to pour down relentlessly. The colors all crouched down in fear, drawing close to another for comfort and safety.

The rain spoke: "You foolish colors, fighting among yourselves, each thinking you are the best. Do you not know that you are all special and that each has a unique and different purpose?"

The rain continued: "Now come with me as I stretch you across the sky in a great bow of color so everyone can see how special you all are! First RED on the outside, then ORANGE, YELLOW, GREEN, BLUE, and VIOLET.

And so, as the rain ended and the sun appeared from behind the storm clouds, a beautiful rainbow arched across the sky, a sign of hope for tomorrow.

based on an Indian legend; written by Anne Hope (1438); adapted (1993)

56. Sunshine to You

Activity

Objective • to create sunny day or rainbow pictures

Materials • white construction paper, water color paint, crayons, markers

Have children create their favorite sunny day or rainbow picture. They can use various forms of art materials either singly or in combination to create these. Since this is a culmination activity, the art should reflect many aspects of what they have learned about weather.

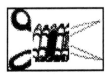

Temperature

57. Human Thermometers Aren't So Hot

Activity

Objectives
• to learn about temperature
• to learn why humans are not good thermometers

Materials
• alcohol-in-glass thermometers, 3 containers for each group, water, ice cubes, unsharpened pencils

Have each child hold a pencil and touch the eraser, metal band, and wooden stem. Ask children which pencil part felt coldest and which felt warmest. Repeat for tables, chairs and other classroom objects that contain metal and non-metal parts. **Did everyone agree?**

The temperature of all parts of the pencil (or other objects) is the same. However, because heat is transferred from the human skin to metal fastest, the children felt this part to be the coldest. Heat transfer from the skin to a non-metal part is the slowest. These parts probably felt warmest to the children.

Group children at several tables. At each, place three containers with water of different temperatures. One container should have room temperature water, one should have warm (but not hot) water, and the third should have water in which several ice cubes have melted. Order the bowls from warmest (left) to coldest (right) at all tables.

Have one child at a time complete the following activity. Place one hand in each of the outer two containers. The child should tell you which container has the warmest water. The child should then place both hands in the middle container. Have the child tell you which hand feels warmest and which coldest. The hand which was "hot" before is now colder, and vice versa. Yet, the temperature of the water in the middle container can't be both warm and cold!

After each child has had a chance to experience this "unusual" sensation, have them measure the temperature of each of the three containers. They should also discuss why humans are not good thermometers.

The children should recognize this from family situations in which one family member felt cold and another felt warm, while both were in the same room. Emphasize "relative" temperature (i.e., warmer, colder) rather than "absolute" readings, especially for younger children.

> LEADER CHALLENGE: Have children go home and take a family poll. **How many family members felt cold or warm in the same room at the same time?** Have the children try this in the evening and again in the morning, if the family schedule permits. Discuss results in class.

58. Reading a Thermometer

Activity

Objectives
• to measure temperature using a thermometer
• to calibrate a thermometer

Materials
• alcohol-in-glass thermometers, containers, water, ice

Show the children how to read a thermometer. Give each child a thermometer (be sure it is labeled with a letter using a permanent marker). Tell them to be careful with the instrument, but that they can gently squeeze the bulb of the thermometer between their fingers. Some may want to blow on their thermometers. **What happens to the red fluid?**

It is important for children to understand that as the fluid inside the thermometer becomes warmer the fluid spreads apart (i.e., expands). The opposite happens when the temperature goes down. Use group movement to have children huddle as they shiver (i.e. get cold). As they warm up, they want to spread their arms and legs apart.

Have each child read the temperature inside the classroom and then outside on the playground each day for a week. They should record their observations on a data sheet. Children can connect the symbols to create a line graph.

Some children may want to measure the temperature of the soil, a puddle, a bowl of ice water, or inside their sleeve. As long as they are careful, they should experiment measuring the temperature of different things. Give them a second data sheet onto which they can place their observations.

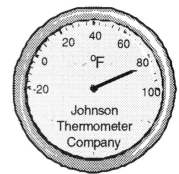

In small groups have children read their thermometers and compare temperatures. **Are the temperatures the same?** Then have children align the thermometers to see what might be different. Some of the children should quickly see that the bulbs do not line up exactly. You can show this to the class by placing three thermometers on an overhead projector. Discuss some other variables that might be causing the difference in temperature?

LEADER CHALLENGE: Ask children to go home and look for thermometers. Remind them that thermometers come in different shapes and can look round and have a pointer like the one shown above. Hint: there are thermostats, cooking thermometers, thermometers for taking one's body temperature, and thermometer gauges on car dashboards. Have children report on the results of their search.

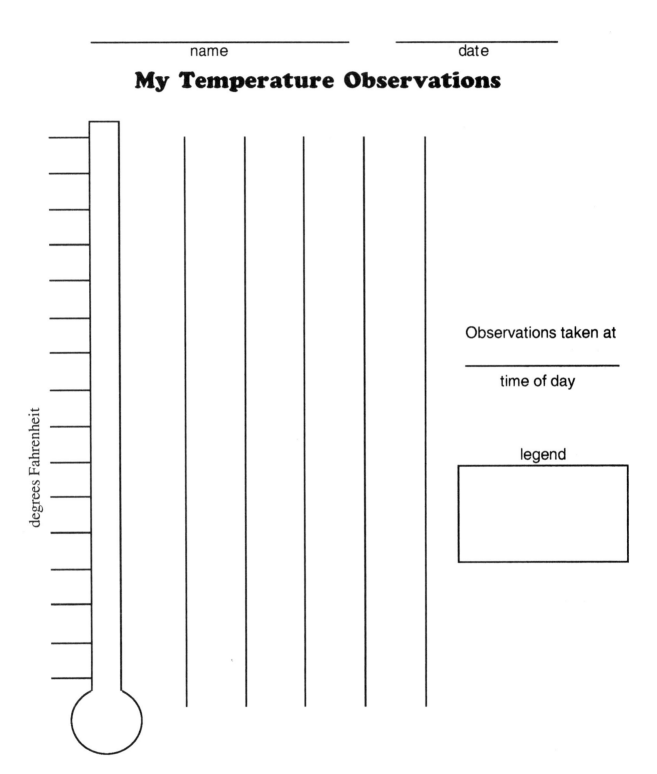

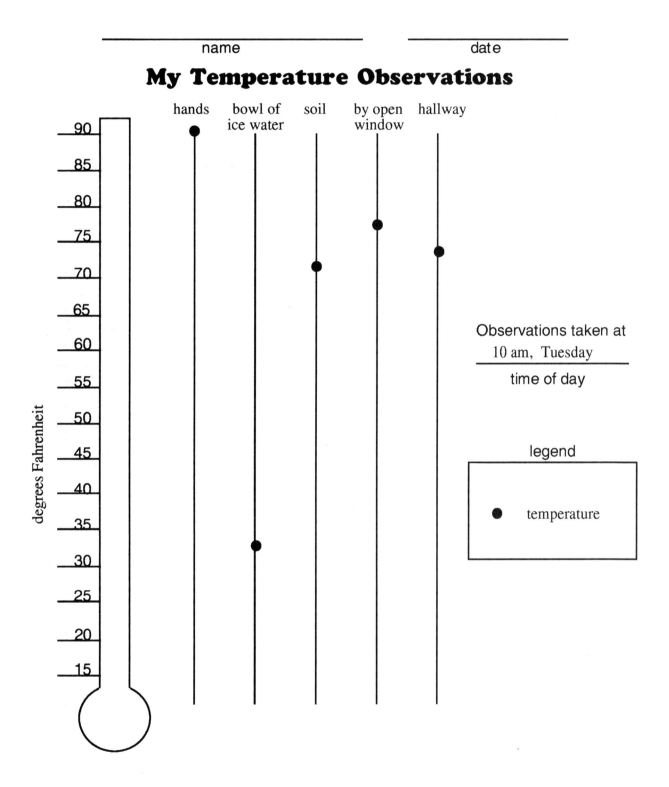

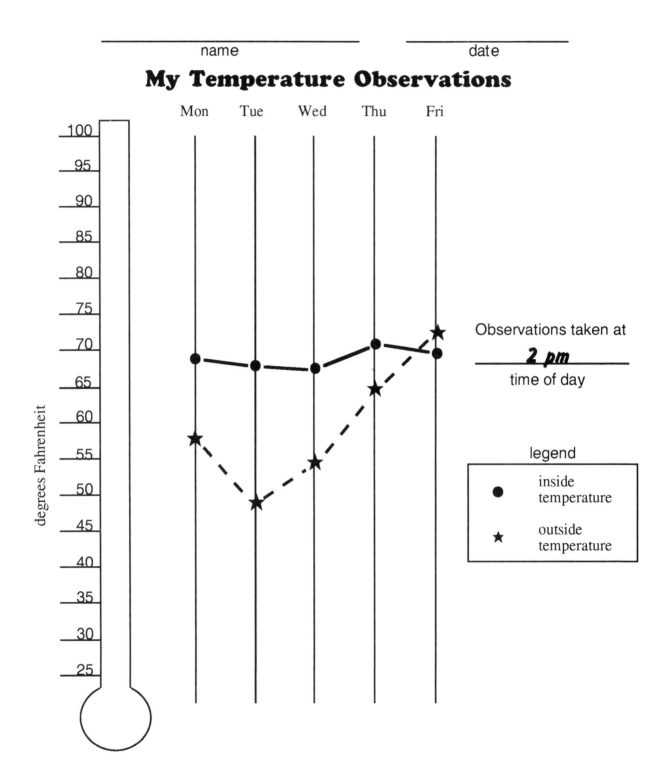

59. Expanding Your Understanding of Temperature

Activity

Objective • to make an air thermometer and use it to measure temperature

Materials • small-mouthed containers, 11" balloons, rubber bands, white glue, flat coffee stirrers, scissors, heat source (e.g., hair dryer) cold source (e.g., ice chest), alcohol-in-glass thermometers

Cut off the neck of an 11" balloon and make a cut from the edge of the balloon to its center. Stretch the balloon over the top of the small-mouthed container. Wrap a rubber band around the neck of the container to hold the balloon in place. Show how teamwork makes this easier to do. Pull the edges of the balloon to stretch it more tightly. Glue the end of half a coffee stirrer onto the balloon, about halfway between the center and the edge. Allow to dry.

Ask children **what will happen to the pointer when the air gets warm? When the air gets cold?** Discuss answers. Demonstrate by using a hair dryer or by placing the thermometer in direct sunlight. Since the air inside the container expands (just like the fluid did inside the thermometer in Activities 57 and 58), the balloon pushes upwards, causing the pointer end of the coffee stirrer to point down. Place the thermometer in an ice chest filled with ice or in a refrigerator to simulate a cold day. This causes the air inside the thermometer to contract, and the pointer end of the coffee stirrer to point up.

Have children measure the temperature inside and outside each day for one to two weeks. Give them an index card labeled with dates or days of the week. Each vertical line starting from the left stands for a new day. Label the card "My Temperature Record" and be sure to place the child's name on the card. Children should be given two cards, one for inside and one outside readings.

Have children stand the index card vertically by the pointer of their thermometer and make a dot where the pointer points each day. Even without a numerical scale, children should be able to see the trace (or graph) of their observations. They can connect the dots (in order) to complete the graph. This is called a

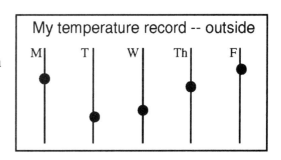

thermograph (temperature graph). Older children can assign a numerical value by also measuring the temperature with an alcohol-in-glass thermometer. This allows them to calibrate the thermometer that they built.

60. Looking for Warm and Cool Places

Experiment

Objectives
• to predict the warmest and coolest places in your playground environment
• to determine how temperatures vary from place to place

Materials
• air thermometers, alcohol-in-glass thermometers, index cards

Have children predict where they think the warmest and coolest places are around their playground. A sunny, cold day will provide the greatest range of temperature. Although they should probably know from experience that sunny places are warmer than shady places, there are other factors which can contribute to temperature differences around the playground.

Divide the class into two teams -- one that is in charge of cool temperatures and one that is in charge of warm temperatures. Children from each team should place both the air (see Activity 59) and the alcohol-in-glass thermometers in the locations that they think would support their team's temperature prediction. Wait 10 minutes and have the children return to their thermometers. They should read the temperature from the alcohol-in-glass thermometer and note the location of the pointer on their air thermometer. It is sufficient to note whether the pointer is pointing up or down a little or a lot and how it has changed during the 10 minute period.

Which places were the warmest? Which were the coolest? Is there anything in common with these groups of places? For example, were the coolest spots all in the shade?

Have children leave their thermometers in a safe place on the playground area throughout the day, periodically checking them. The older children can record the temperatures each hour; the younger children can just note whether the temperature got warmer or colder (the red fluid in the thermometer changes or the pointer changes on their air thermometer). Children can repeat the thermograph data collection each hour instead of each day as described in Activity 59. Children should find that thermometers in sunny areas change more than those in shady areas.

> LEADER CHALLENGE: Have children take their plastic jar thermometer home. They should place it inside and outside of their home and record relative temperatures on another index card. With parental supervision, they should place the thermometer in their refrigerator and freezer and add these values to their index card.

  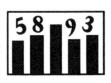

61. Things Are Heating Up

> **Activity**

Objective • to predict which color heats up fastest in sunlight

Materials • assorted colored construction paper

Show the children between five and eight pieces of construction paper, each a different color (be sure to include black and white). Ask children to predict which color heats up fastest in sunlight. They should line up in front of their prediction. Make a bar graph of this human prediction graph and post it in the room. Give each child a piece of the colored paper they chose. Repeat for the color that they think heats up the slowest, making a separate bar graph. Each child will have two different colors of paper.

Each child takes their papers outside and places them in a sunny area. Weight down the papers with a heavy object, if needed. Return in about 5 minutes. Have each child touch their two pieces of paper and note which one is warmer. Pick up the papers and go inside.

Have each child report their findings. You can mark W (for warmest) and C (for coolest) on the two graphs you have posted. If one child had predicted that red would be warmest, and that child reports that red was the warmest, then place a W in a red box on the warmest graph. Plot accordingly for cool predictions and observations. The more W's on the warmest graph and the more C's on the coolest graph, the more accurate the predictions.

Which color(s) were the warmest? Which were the coolest? What color clothes do you wear in the summer? In the winter? Do you think that the color of the clothes you wear helps you feel warmer or cooler?

Use the book *People* by Peter Spier to explore differences in clothing and housing among peoples from around the world.

To quantify the experiment, you may wish to wrap a thermometer completely in the each of the various colored construction papers. Place in a sunny location and compare temperatures after 10 minutes.

Suggested reading: *People*

62. The Heat is on

Activity

Objective • to determine which ground surface heats up fastest in sunlight

Materials • none

Have children explore the area around your school. Have them touch different objects on a sunny day to determine which ones heat up the most in sunlight. Be sure they touch playground equipment, the siding of your building, grass or plant leaves, soil, concrete or asphalt, and fences. *Be careful! Some objects (particularly those made of metal), can be very hot, especially on sunny, warm days!!*

Which surfaces were the warmest? Which were the coldest? Aren't you glad that your playground equipment does not get hot when the sun shines on it? Or does it get hot?

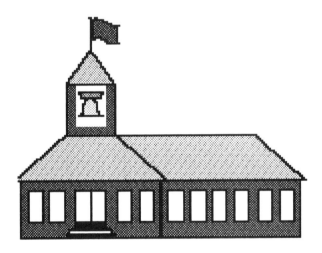

63. Now You See It - Now You Don't

Activity

Objective • to predict how fast an ice cube will melt in any season in both sunny and shady locations

Materials • ice cubes, thermometers, styrofoam trays

Have children measure the outside temperature and then predict how long it will take to melt their ice cubes in a sunny location and in a shady location. Have children record their predictions. Each child should take two ice cubes and place each on a styrofoam tray. They should place one tray in a shady location and the other in a nearby sunny location. Note the time that the ice cubes and trays are placed outside.

Check on the ice cubes each hour on a cold day and every 10 to 15 minutes on a warm day. When did the ice cubes finally melt? Record data and discuss. Repeat the experiment on a day when the outside temperature is significantly different.

Older children can predict the time at which they think the ice cubes will melt and subtract to find the amount of time the process took.

Which ice cubes (sunny or shady) melted fastest? Did the ice cubes melt fastest on a warm or cold day? If the temperature is below freezing will the ice cube melt? What happens to the ice cube in the shade on a day when the temperature is below freezing?

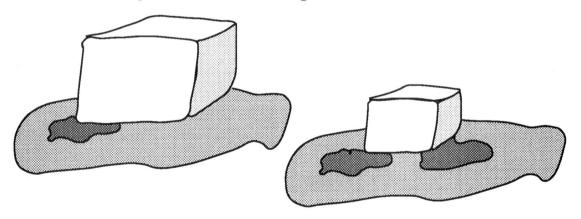

LEADER CHALLENGE: See Activity 25. Have children place an ice cube in their freezer at home. This is similar to a shady area on a day when the temperature is below freezing. Have them observe what happens to the ice cube over several days.

_____ _____
     name   date

Melting Predictions and Observations

Today is a   warm day   cold day

I placed my ice cubes in sunny and shady places at _____
and predicted how long it would take them to melt.   time

	Sunny location	Shady location
Predicted time of melting		
Actual time of melting		
Number of minutes until melting		

How close were your predictions? _____
_____
_____
_____
_____
_____

© 1996...H. Michael Mogil and Barbara G. Levine, 1522 Baylor Avenue, Rockville, MD 20850

64. Putting Chemistry on Ice

| Activity |

Objectives
- to explore temperature change through chemical interactions
- to brainstorm ways that salt is used to melt ice

Materials
- dissolvable substances (e.g., salt, sugar, corn starch, baking soda, alum), ice cubes, water, craft sticks, plastic cups, alcohol-in-glass thermometers

Place children in groups of six and give each child an alcohol-in-glass thermometer. Provide three children plastic cups filled halfway with water and three with cups filled halfway with ice cubes. Have children measure the temperature of the contents of each cup. Record this data and data to be collected in the next steps on chart paper.

Place a small amount of sugar into one ice and water cup. Do the same for salt and at least one other substance. Have the children gently stir the contents of their cup. After about a minute, have them read the temperature. **Did it change?** Continue stirring gently for about three minutes. **In which of the cups was the largest temperature change observed?**

Have children discuss situations in which salt is applied to ice. **What happens to the ice when salt is placed on it?** If you live in an area in which snow and ice storms do not occur, you may need to explain to the children how salt is used to melt snow and ice on sidewalks and highways. Salt is also used to make ice cream and ice milk. See Activity 65.

65. We Get to Eat an Experiment!!!

Activity

Objectives
- to make ice milk (and eat it)
- to measure the temperature of the salt-ice mixture that freezes the ice milk

Materials
- table salt, milk, ice cubes, food flavorings (e.g., vanilla, chocolate), sugar, craft sticks, soup or similar metal cans, paper cups, empty plastic gallon-size containers (e.g., ice cream or sweater box), measurers (cup, tablespoon, and teaspoon), oven mitts

To limit the amount of materials needed for this activity, we suggest having children work in pairs. This is especially important for younger children who may become tired of stirring during this activity.

Prepare the gallon container. Layer ice cubes and salt so that the container is about three-fourths full.

Working with each pair of children, pour 1/2 cup of milk and one tablespoon of sugar into a soup can. Place the can into the center of the gallon container. and gently twist it into the ice so that most of the can is submerged. Place a thermometer in the ice-salt mixture (be careful not to force it in; thermometers can easily break), and another on the table nearby. Be careful that children do not touch the ice-salt mixture or the outside of the plastic container. This mixture is very cold, as you and your children will soon discover.

Give each pair a craft stick and have them take turns gently stirring the milk mixture. As ice starts to form on the inside of the can, be sure to scrape it away and mix it with the milk mixture. It should take about 15 - 20 minutes for the milk mixture to freeze. Then add one or more teaspoon of flavoring and stir. Take the ice milk from each soup can and place into paper cups. Allow the children to eat their creation. MMMMM!!!

ice cream
- mix milk and sugar in a metal can
- place can in a larger container and surround can with ice-salt mixture
- stir milk, for 15 minutes or more
- add more ice to ice-salt mixture
- stir until frozen
- add flavorings and enjoy!!

While the children are eating their ice milk, have them gather around their two thermometers. Have children read the temperature of the ice-salt mixture and compare this temperature to the temperature of the room. Discuss their observations.

Children should notice that ice forms on the outside on the plastic ice cream container. This is like dew, except that it is frozen. It is called frost. Frost forms when condensation occurs at temperatures below freezing. **Have children seen frost on cars, roofs, or on grassy areas?**

An alternative activity would involve using different types of salts (e.g., Kosher, rock, table). **Which one chills the ice-salt mixture the most?**

When the ice completely melts, carefully pour off water from the ice-salt mixture, using oven mitts to hold the plastic container. Discard any remaining rock salt in a plastic bag; the other salts may be washed down a drain without clogging it.

Note: You can substitute skim for regular milk, and sugar substitute for sugar, based on any food allergies children may have. You can also freeze fruit juices in the same manner.

> LEADER CHALLENGE: Have children bring home these instructions for making ice milk. Encourage them to make ice milk with their families. Instead of a craft stick, they should use a wooden spoon. Have children brainstorm why wooden stirrers are used for this activity.

66. Blowing Hot and Cold Bubbles

Activity

Objectives
• to make a bubble mixture
• to study how temperature affects floating bubbles

Materials
• water, Dawn™ dishwashing detergent, glycerin, pipe cleaners, bubble pans and/or cookie sheets, straws, string

Create bubble wands using pipe cleaners or thread string through straws to create giant bubble makers. For pipe cleaners, have children bend one end into any shape they want and twist to make a complete enclosed shape, or tie several together to create more complex shapes.

Mix one gallon of very warm water, two-thirds of a cup of Dawn™ dishwashing detergent, and one tablespoon of glycerin to create a "warm" bubble mixture: repeat using cold water to create a "cold" bubble mixture. Do not tell children which is which.

Pour some "warm" mixture into one bubble pan/cookie sheet, and some cold into the other. Have children blow bubbles. What do they observe?

Have children brainstorm to determine why some bubbles floated up, while others fell to the ground.

Although the activity works best on a cold day, we suggest that you try it on days with different temperatures. This may be during the afternoon (when the temperature outside warms up), on another day the same week, or even in a completely different season.

Refer to Activity 58 for related activities that explain how cold temperatures cause air to contract and warm temperatures cause air to expand.

Also have children examine their bubbles for rainbow patterns (Activity 54).

 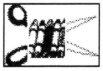

67. 'Tis the Season

Activity

Objective • to compare and contrast how children experience seasons

Materials • large sheets of drawing paper, markers, crayons, paint

Give each child a large (12" x 18") piece of drawing paper. Have them fold it into fourths (6" x 9" sections). Label each quarter with a seasonal name and have each child draw or paint themselves doing a seasonal activity. It should include proper clothing (hopefully the proper color), proper weather backdrop, and show them having fun.

Discuss with children their seasonal experiences. **Do they only know the four standard seasons? Have they ever experienced "holiday" seasons, vacation seasons, wet or dry seasons, or a fire season? Have they ever seen a weather event happen in the "wrong" season (e.g., snow during the spring or a thunderstorm in winter)?**

Make a human graph of birthdays by season. **How many children were born in each season?** Be careful. We know a boy who was born during the fall, but thinks he was born in the winter. The reason is that it snowed on the day he was born.

Suggested reading: *First Comes Spring*

 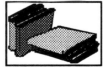 

_____

is recognized as an amateur meteorologist

for having successfully studied about

- **Sky**
- **Water**
- **Wind**
- **Sun**
- **Temperature**

_____
teacher

_____   _____
school / center   date

_____
meteorological development team

© 1996...H. Michael Mogil and Barbara G. Levine, 1522 Baylor Avenue, Rockville, MD 20850

References

Selected Reading Books linked to activities

Branley, Franklyn M. (1988) <u>Tornado Alert</u>. New York, NY, Harper-Collins
Branley, Franklyn M. (1961) <u>What Makes Day and Night</u>. New York, NY, Harper-Collins
de Paola, Tomie (1975) <u>The Cloud Book</u>. New York, NY, Holiday House
Ets, Marie Hall (1963) <u>Gilberto and the Wind</u>. New York, NY, Penguin Books
Ginsberg, Mirra (1974) <u>Mushroom in the Rain</u>. New York, NY, Aladdin Books
Keats, Ezra Jack (1968) <u>A Letter to Amy</u>. New York, NY, Penguin Books
Keats, Ezra Jack (1962) <u>The Snowy Day</u>. New York, NY, Penguin Books
Koscielniak, Bruce (1995) <u>Geoffrey Groundhog Predicts The Weather</u>. New York, NY Houghton Mifflin
Rockwell, Anne (1985) <u>First Comes Spring</u>. New York, NY, Harper-Collins
Shaw, Charles, G. (1947) <u>It Looked Like Spilt Milk</u>. New York, NY, Harper-Collins
Showers, Paul (1991) <u>The Listening Walk</u>. New York, NY, Harper-Collins
Spier, Peter (1980) <u>People</u>. New York, NY, Doubleday

Other Reading Books

Alexander, Martha (1992) <u>Where Does The Sky End, Grandpa?</u> New York, NY, Harcourt Brace Jovanovich
Bauer, Caroline, editor (1986) <u>Snowy Day: Stories and Poems</u>. New York, NY, Harper-Collins
Branley, Franklyn (1986) <u>Air Is All Around You</u>. New York, NY, Harper-Collins
Briggs, Raymond (1990) <u>The Snowman Storybook</u>. New York, NY, Random House
Felix, Monique (1993) <u>The Wind</u>. Columbus, OH, American Education Publishing
Freeman, Don (1966) <u>A Rainbow of my Own</u>. New York, NY, Penguin Books
Freier, George (1989) <u>Weather Proverbs</u>. Tucson, AZ, Fisher Books
Frost, Robert (1978) <u>Stopping By Woods on a Snowy Evening</u>. New York, NY, Dutton Books
Fuchs, Diane (1995) <u>A Bear for All Seasons</u>. New York, NY, Henry Holt and Company
Groth-Fleming, Candace (1994) <u>Professor Fergus Fahrenheit and his Wonderful Weather Machine</u>. New York, NY, Simon and Schuster
Mayper, Monica (1991) <u>Oh Snow</u>. New York, NY, Harper-Collins
Miller, Edna (1990) <u>Mousekin's Frosty Friend</u>. New York, NY, Simon and Schuster
Reay, Joanne (1995) <u>Bumpa Rumpus and The Rainy Day</u>. New York, NY Houghton Mifflin
Rogasky, Barbara (1994) <u>Winter Poems</u>. New York, NY, Scholastic Inc.
Silver, Norman (1995) <u>Cloud Nine</u>. New York, NY Houghton Mifflin
Yolen, Jane (1987) <u>Owl Moon</u>. New York, NY, Philomel Books

Other References / Resources

Bianchi, John and Frank Edwards (1992) <u>Snow</u>. Newburgh, Ontario, Canada, Bungalo Books

Branley, Franklyn (1986) <u>Sunshine Makes the Seasons</u>. New York, NY, Harper-Collins

Caduto, Michael J. and Joseph Bruchac (1988) <u>Keepers of the Earth</u>. Golden, CO, Fulcrum

Copycat Magazine (March/April 1995, Volume 10, Number 4). Racine, WI, Copycat Press

Freier, George (1989) <u>Weather Proverbs</u>. Tucson, AZ, Fisher Books

Hosking, Wayne (1990) <u>Flights of Imagination</u>. Washington, DC, National Science Teachers Association

Locker, Thomas (1995) <u>Sky Tree</u>. New York, NY, Harper-Collins

Prelustky, Jack (1983) <u>The Random House Book of Poetry for Children</u>. New York, NY, Random House

Sherwood, Elizabeth *at al* (1990) <u>More Mudpies to Magnets</u>. Mt. Rainier, MD, Gryphon House